SCIENTIST versus SOCIETY

SCIENTIST VERSUS SOCIETY

Vivian Werner

HAWTHORN BOOKS, INC.
Publishers/NEW YORK

ACKNOWLEDGMENTS

I wish to acknowledge my everlasting debt to Dr. Samuel Wolfenstein, Professeur de Mathematiques at the Faculte des Sciences at Le Mans, for his numerous useful suggestions, his help and understanding, and above all for his great patience in explaining the fundamentals of mathematics, which made the work of many of the subjects of this book comprehensible to me and, I hope, to my readers as well.

Paris, 1975

SCIENTIST VERSUS SOCIETY

Copyright © 1975 by Vivian Werner. Copyright under International and Pan-American Copyright Conventions. All rights reserved, including the right to reproduce this book or portions thereof in any form, except for the inclusion of brief quotations in a review. All inquiries should be addressed to Hawthorn Books, Inc., 260 Madison Avenue, New York, New York 10016. This book was manufactured in the United States of America and published simultaneously in Canada by Prentice-Hall of Canada, Limited, 1870 Birchmount Road, Scarborough, Ontario.

Library of Congress Catalog Card Number: 74-33589
ISBN: 0-8015-6586-3
1 2 3 4 5 6 7 8 9 10

To Dale and Cameron

CONTENTS

I	Modern-Day Martyrs	*11*
II	Charles Babbage	*19*
III	Ada Augusta, the Countess of Lovelace	*42*
IV	Gregor Mendel	*61*
V	Sigmund Freud	*87*
VI	Charles Drew	*114*

VII   Nicolai Ivanovich Vavilov   *136*

Bibliography   *151*

Index   *155*

SCIENTIST versus SOCIETY

1
MODERN-DAY MARTYRS

IN THE YEAR 1859, a bombshell in the form of a book burst upon the world. It was called *The Origin of Species by Means of Natural Selection*, and its author was a shy, middle-aged English naturalist named Charles Darwin.

He had spent nearly half his life working out the theory he now set forth. It held that all species of plants and animals developed from earlier forms. Such an evolution took place, he said, when slight hereditary differences were transmitted from one generation to the next. When such a difference gave an advantage to the plant or animal in its struggle for existence, it survived, to pass along the new or modified characteristics to its descendants. Darwin described this process as "the survival of the fittest," a phrase that is still generally accepted.

Darwin's carefully worked out thesis, with his five hundred pages of clearly reasoned thought and his myriad examples, attempted to answer two questions that had bewildered human beings in even the most primitive societies. The first concerned humans themselves: Where did they come from? The second concerned all of nature: How could its infinite variety be explained?

The idea of evolution had been considered in the century or so before Darwin's brilliant explanation shook society. His own grandfather, Erasmus Darwin, had himself written a book on the subject. That work closely followed the theory advanced by the French naturalist Jean Baptiste Pierre Antoine de Monet, Chevalier de Lamarck, who believed that acquired characteristics could be inherited. As an example, he cited the giraffe which, he thought, stretched its neck to reach the leaves on the highest branches of the trees in the jungle. Its offspring, therefore, would be born with long necks, enabling them, too, to feast from the top of the forest. Had Lamarck been asked whether or not the leopard could change its spots, he surely would have answered, "Yes."

Lamarck's views attracted some attention during his lifetime and even today influence certain scientists. But they had nothing like the impact of Darwin's theories. For the most part, men clung to the idea that every living thing was created exactly in its present form. The debate that developed around Lamarck's ideas was part of the age-old conflict between science and religion. And Lamarck was not to resolve it.

Religion—or faith—had always served to dispel the terrors of the unknown. For early people, those terrors were legion. Fire was a mystery, completely incomprehensible. Birth and death—life itself—were beyond understanding. Thus, it was essential that a supernatural explanation be found. Surely there was some all-powerful being—or many powerful beings—who controlled the setting of the sun and the rising of the moon, the ebb and flow of the tides, who unleashed the terrible storms that destroyed crops. Therefore, there must be some sort of deity, some being of a divine nature.

Modern-Day Martyrs

Early people believed in one or many. At first their gods were those they could see. They worshiped the sun and the moon; they worshiped the stars. From time to time they worshiped animals. And sometimes they attributed miraculous powers to the objects around them. Was there not some form of magic in the rocks and in the trees? It seemed so. And since it seemed so, why not worship them?

But deities, it soon appeared, could be brought closer, kept directly at one's side, if they took the form of idols carved of wood or stone. Almost always, those idols were made in the human image. Inevitably, they were thought to possess supernatural powers. Those who were pleased would guard the persons who had sculpted them. And those who were offended would wreak the most horrible vengeance.

Humans turned to their gods for explanations whenever they were confused. There was no need for knowledge or study; faith alone was enough. And that faith provided the structure of life, the framework within which it was lived.

Even then, though, there were some who sought other reasons, who turned from faith to the rational. Little by little they enlightened the world, rolling back the darkness of superstition. Those men were the very first scientists. Undoubtedly they, too, were persecuted for their advanced notions. It has never been easy to substitute a new idea for one that has been accepted without reservation.

Gradually, however, they *were* accepted, and as people's knowledge increased, their world expanded. Superstitions still abounded, as they do today in many parts of the world. But in both East and West, many of them disappeared. Cultures developed and barbarism receded. Civilizations, in the forms we now know, were established, growing slowly but surely, eventually spreading over much of the face of the earth.

Scientist Versus Society

One of the major concepts of the new civilizations was that of monotheism, the belief that there was only one God, rather than the many who had earlier been supposed to control all beings, all events. In the East, the idea was accepted by the Buddhists and the Moslems. In the West, Judaism represented this point of view. And from Judaism came Christianity, embodying most of the Judaic concepts and traditions.

With monotheism appeared a new vision of God. It was just the reverse of that held in primitive societies, where the gods were like men. Now it was man who was like God—"created in his image." Since this was presumed to be so, and since God was both omniscient and omnipotent, that which he had created in his image must inevitably be the center of the universe.

The idea was strictly adhered to by the Roman Catholic Church, which emerged, after the Middle Ages, as the most powerful institution in the West. Humans, for all their progress, had not lost their need for faith. It was in the church that they found it. As before, religion provided the answers to questions that could not otherwise be answered. The church, therefore, held society together, once again establishing the standards by which people lived, the framework of their lives.

As early as the sixteenth century there were savants—scientists—whose own observations of the world around them caused them to dispute the teachings of the church. Among them was the Polish astronomer Copernicus, who used the newly invented telescope to make vital discoveries concerning the earth itself. Until then, it had been assumed that the sun, along with the few known planets, revolved around the earth. But Copernicus showed, by studying the stars and making careful calculations, that the earth was part of a solar system that included many planets. And all those planets, he proved, revolved around the sun. It must inevitably follow, then, that humans were not the center of the universe, since the earth itself was not.

Such a revolutionary idea—the exact opposite of what was "known"—was pure heresy, and as such was bound to bring down the wrath of the church authorities on the heads of those who held it. Copernicus himself was fortunate; his work was not published until he was on his deathbed. Others, like the Spanish physician and theologian Servetus, were burned at the stake.

The most famous defender of the theory of Copernicus was the great Italian astronomer Galileo. He proved, beyond the shadow of a doubt, the validity of the ideas set forth by his illustrious predecessor. And he clung to his beliefs, despite the opposition of the church. But when he was tried by a group of the highest church officials and threatened with the same fate as befell Servetus, he recanted. As a result, his life was spared, but he spent the rest of his days under house arrest in the city of Florence.

While religion and religious institutions were still of supreme importance, there already were important divisions in the Catholic Church. Martin Luther, in Germany, had broken away and set up his own sect. In England, the king himself had refused to accept many of the laws handed down by the Pope in Rome, and a new church—the Church of England—was in the making.

Scientists now had more freedom. In Germany, Johannes Kepler expanded the original ideas of Copernicus. Sir Isaac Newton, the English mathematician, pursued much the same course without interference. His ideas, in fact, served as the basis of scientific thought until recent times.

The Protestant Church lacked the vast powers—the real

Scientist Versus Society

authority—that the Catholic Church had exercised in earlier times. Nevertheless, it exerted a moral influence on its members that was rarely ignored.

The absolute authority of the Catholic Church over secular matters in countries where it was still dominant had also waned. It retained much of its former control, however, shaping both lives and thought as before. Whatever their respective weaknesses, religious institutions throughout the Western world continued to set the standards to which most individuals conformed.

At the time of the publication of *The Origin of Species,* belief in the teachings of the Bible as the literal truth, especially as applied to the story of creation, was almost universal. Geologists attempted to prove this concept by what was known as the catastrophic theory. They began by noting the formation of rocks and by examining the fossils—the remains of animals and plants—that were found at different strata of the earth. Each stratum represented a different time period; each time period represented hundreds of thousands of years. They came to the conclusion that at various times God had unleashed furious—or catastrophic—forces upon the world, wiping out all living things. Later he had created new and completely different species that replaced the earlier ones.

It was Darwin's book that dispelled that doctrine. His arguments were convincing, his reasoning was logical and forceful. Yet he was assailed by many.

The great Swiss naturalist Louis Agassiz, the first foreigner to become a professor at Harvard University, denounced the work, stating, "the facts about living things proclaim aloud the one God, whom man may know, adore, and love." The clergy, as was to be expected, was especially rabid in denunciations of Charles Darwin and his "atheistic" notions.

Modern-Day Martyrs

In his defense, however, Darwin could count on many of the outstanding scientists of his day. The furor over *The Origin of Species* quickly died down, at least among the well-educated. But there were—and are to this day—pockets of resistance to the idea of evolution. Between 1921 and 1929, bills prohibiting the teaching of evolution were introduced in twenty states in the United States. In 1925, in a historic trial in Tennessee, a twenty-four-year-old science teacher named John Thomas Scopes was found guilty of the crime of teaching the Darwinian principles of evolution in his biology class.

In general, however, Darwin's theory has been universally accepted. He had, indeed, succeeded in establishing it as a scientific fact.

Darwin had shown that people were descended from earlier forms of primates—the most highly developed species of animal life, which includes monkeys and apes, and which in turn had descended from the lower and even the lowliest forms of life. In doing so, he had put an end to the idea that humans were created in the image of God. Actually he had gone much farther; he had denied the nature of the God in which people had believed for so long.

Suddenly the entire structure on which humanity had depended crumbled away. There was no longer the Almighty on whom all responsibility rested. That responsibility had been shifted to the shoulders of each and every individual.

With religion—and the church—no longer providing the form for living, it was essential to seek a different one. This was found in society itself, in the ideas and attitudes that prevailed in the community. In a fine example of poetic justice, one of the new ideas that held sway was based on a Darwinian principle, the survival of the fittest. In the industrial expansion of the late nineteenth century, the

17

Scientist Versus Society

acquisition of great fortunes through the exploitation of workers by unscrupulous employers was handily excused on the grounds that this was in nature's scheme of things.

A new era had come into being with the publication of Charles Darwin's masterpiece. With society now dominating life as the church had before, the old conflict between science and religion was ended. But new conflicts—between science and society—replaced the old.

One of the most original of all scientists, Charles Babbage, offended many of his countrymen because he refused to accept the English class system, believing that science was a profession and not simply the pastime of gentlemen. His friend Ada Augusta, the countess of Lovelace, herself a mathematical genius, was never allowed to develop her great talents because she was a woman.

Gregor Mendel was completely ignored; his ideas supposedly contradicted those of Darwin, which had by then been accepted. Sigmund Freud was suspect from the first because he was Jewish in a Catholic country. But his greatest offense was offending Victorian morality by theories about the sexual basis of emotional disturbances at a time when the mere mention of sex was taboo.

Charles Drew was black in a white world. And his work, which proved that people were biologically the same, was utterly unacceptable. Finally, an entire generation of Russian geneticists was punished for refusing to accept the discredited Lamarckian theory, which was being used for the purpose of propaganda by the new Communist rulers of their country.

Here are the stories of these men—and this woman—who became modern martyrs of science.

II
CHARLES BABBAGE

IT WAS A noble gesture on the part of Charles II, king of England, to grant a charter to the group of gentlemen whose interest in science was so great that they met regularly to discuss the latest advances in the field. In an even nobler gesture, in 1662 King Charles gave the group the right to call itself the Royal Society. Under the patronage of the reigning monarch, its members would be encouraged to explore new ideas and exchange views. Those talented men of the upper classes would certainly find in the company of their peers the intellectual stimulation that could only come from free and open discussion.

The idea of such a society was not new. In France, such great thinkers as Pascal and Descartes were already meeting regularly. By 1666, the celebrated statesman Colbert had offered them the use of the royal library for their assemblies. By the end of the century, they were gathering in the enormous palace in the heart of Paris, the Louvre. Now the society was given the official title of the Académie des Sciences.

Scientist Versus Society

Moreover, its expenses, although minor, were met by the state. The Académie continued to meet until revolution swept through France in 1789. It was suppressed for a short time during those troubled days, but within a few years it was again functioning, as it does to this day.

The revolution in France brought with it new attitudes and customs, making many of the old ones obsolete. The new government realized the great need for scientific research and understood that it must be supported by the state. Accordingly, ministries were set up to direct projects with the aim of establishing fundamental laws. Under their auspices, men were hired and paid to work out a system of standard weights and measures, far more accurate than any previously known. Others computed mathematical tables such as logarithms, which were indispensable to the navigators of ships then roaming the seas with increasing frequency, carrying their cargoes to distant parts of the world.

The very concept of research as the responsibility of the government was a radical innovation. But equally radical was the decision to employ any person of talent, regardless of social standing. The motto of the new French Republic was *Liberté, Egalité, Fraternité*—Liberty, Equality, Fraternity. And the new and forward-looking leaders were determined that those ideals were to be put into practice. Here, in the new bureaus, was the opportunity to do so.

Such an idea was almost unthinkable in England, just across the Channel. Gentlemen were gentlemen and science was a hobby for the upper classes.

Most of the gentleman scientists had been—and would be—educated at either Oxford University or Cambridge. Both were renowned for the classical studies they offered—Greek and Latin as well as history and philosophy. It was true that the teaching of science lagged far behind that in the great

Charles Babbage

universities on the Continent. However, few students seemed to notice. But to young Charles Babbage, entering Trinity College at Cambridge, where the great Sir Isaac Newton had once held a chair, the knowledge was a shock.

Babbage had been a sickly youth almost from the time of his birth in the tiny town of Totnes, Devonshire, on the day after Christmas in 1792. Consequently, he had had little formal education. But he had had an overwhelming interest in anything that smacked of the mechanical. Even as a tiny boy he had asked about each toy given to him, "What makes it work?" Often he broke open his toys to find the coils and springs that sent the coach-and-four spurting forward, the wheels that set the tiny clock ticking. Later he had worked with a tutor from Oxford University, studying mathematics. When he at last entered Cambridge in 1811, at the rather advanced age of nineteen, he was disheartened to realize that he knew more than his don, or tutor.

Nevertheless, he found much at Cambridge to compensate for the sterility of the academic work. Life at the university was agreeable and there was nothing more pleasant than to cut classes—after sending his servant with a certificate from the local pharmacist saying that he was too ill to venture out—and sail down the river Cam with a group of friends, enjoying the hamper of food and the bottles of fine claret they carried with them. In the lovely town itself, with the gothic buildings of Cambridge towering above sweeping lawns, Babbage found welcome companions. Many of the young men he knew were destined for fame; all were alert and intelligent. With those who were closest to him, Babbage—in his second year at the university—formed a society dedicated to the development of mathematics. They called it the Analytical Society, and all members vowed to "do their best to leave the world wiser than when we found it."

Scientist Versus Society

The youths hired a room just off the university campus where they could wander in and out, meet their friends, discuss the important new scientific developments of the day, or simply sit and study whatever interested them most.

Babbage was sitting there one day—the year was 1812—poring over a table of logarithms. People's very lives depended on their accuracy, yet Charles noted one mistake after another.

He looked up as a friend sauntered past. "You know," he mused, "I've been thinking that all these tables could be calculated by machinery."

The tables Babbage mentioned happened to be logarithms. But he might just as well have been referring to trigonometrical ones, or any of the myriad others that must be computed for the use of surveyors, architects, and engineers, as well as astronomers and navigators.

All the tables Babbage had before him had been worked out, with infinite patience and labor, by men. And men, unfortunately, are all too often fallible. Those who made the original calculations made mistakes, something both normal and inevitable. But when the tables were printed, printers' errors crept in. Later tables had been printed from tables that already were inaccurate, thereby compounding the faults. Surely, Babbage reasoned, if the computations could be done mechanically, human error could be avoided.

Yet the young student spoke at a time when there was scarcely any machinery available. Steam had recently been harnessed, but ships were still propelled by the wind that filled their sails. Electricity was understood by only a handful of the more advanced thinkers, and the uses to which it later would be put were beyond the realm of even *their* dreams. The machine that Charles Babbage had in mind—part of which he would eventually build—would most certainly have been incomprehensible to them.

Even before Babbage, others had tried their hands at machines meant to calculate tables and relieve men of such drudgery. The first was built by the French mathematician and philosopher Blaise Pascal in 1642. Three decades later, in 1673, the equally eminent German, Baron Gottfried Wilhelm von Leibnitz, also a mathematician and philosopher, constructed a working model of another such machine. Yet the machines of Pascal and Leibnitz were slow as well as inaccurate, and therefore highly impractical. Babbage promised himself that his own machine would be far different and far more successful than the preceding ones.

But he was not yet ready to start work on it. The days at Cambridge were hardly long enough for the pursuit of all the young man's interests as it was. He spent much time in study, of course, since he intended to be first in his class. Competition from other students—notably Charles's friends John Herschel and George Peacock, both destined to become important scientists—was brisk. One or the other, Babbage felt, would surely carry off the honors at his own college, Trinity. In that case, why not transfer to Peterhouse, another Cambridge college, where his chances of being first—of being named "senior wrangler"—would be better?

Whether or not Babbage would have outshone his brilliant companions at Trinity Hall is a matter of conjecture; in any case, the young man transferred to Peterhouse and fulfilled his intentions. He was indeed first in his new college.

Mathematics, while Babbage's first and lasting love, was not his only one. Chemistry fascinated him and he set up a small laboratory of his own where he carried out complicated experiments.

Scientist Versus Society

While Charles was happy to devote much of his time to such serious matters, he enjoyed more frivolous pastimes, too. Like many mathematicians both before his time and after, he enjoyed playing cards. Consequently, he joined the Whist Club at Cambridge. He also was curious about the supernatural. As a small boy, he had even made a pact with the devil, destined to prove the latter's existence. When Satan failed to materialize, thus fulfilling his part of the bargain the youngster had struck with him, Charles completely lost his religious fervor. It was only normal, under such circumstances, that he should now join the Ghost Club.

He swam in the Cam and took time off for long trips along that river. Moreover, he began his courtship of the lovely Georgiane Whitmore, a young woman from a wealthy and well-born family that boasted of distant connections with the English nobility. In 1814, just after Babbage took his Cambridge degree, he and Georgiane were married.

Young Charles already had published several papers of scientific importance before he left Cambridge. Now he was ready to return there, accompanied by his wife, for further study. Although Georgiane was expecting the couple's first child, there was no need for Babbage to go to work to support his growing family. His father, Benjamin Babbage, was a highly successful banker, known almost affectionately as Old Five Percent because he was one of the first to lend money at that rate of interest. There was money to spare in the Babbage family, as there was in the Whitmore family. So the young husband and wife took up residence in Cambridge while Charles pursued his academic career and once again found time to write scientific papers. These proved so brilliant that in 1816, while still a graduate student, Babbage was elected to the Royal Society. Moreover, he and his friends from the Analytical Society found time to translate a work on calculus,

Differential and Integral Calculus, by a renowned French mathematician, Silvestre François Lacroix, adding two volumes of examples to the original work.

All this brought prestige to the youthful genius. But it proved a handicap, too, for the positions he took, including his support for Lacroix, were almost diametrically opposed to those taken by the leading English mathematicians.

The latter had been greatly influenced by Sir Isaac Newton, the discoverer of the principle of gravity, and they were loath to discard or modify any of his ideas.

One of Newton's contributions to scientific thought had been his formulation of the elements of differential calculus—exactly the subject that had been taken up later by Lacroix, along with Leibnitz. Those two had found a newer and better system of notation, which they quickly put to use. But the English scholars, faithful to their own, refused to accept it. They stuck to the "dot" system of Sir Isaac.

While still an undergraduate, Babbage—along with his friends Herschel and Peacock—had written a devastating paper on the subject. They were all in favor, they said, of the D-ism of Leibnitz, as opposed to the Dot-age of Cambridge University. The double pun—"deism" referred to a philosophy of reason that proved the existence of God, while "dotage" of course meant the childishness of old age—provoked as much indignation as laughter. The laughter was soon forgotten, but the indignation remained to haunt Charles Babbage. He had attacked the scientific establishment, and its members were not about to forget it.

Babbage first felt their fury when he applied for a job as professor of mathematics at East India College in Haileybury. In spite of his remarkable qualifications, the post went to another—and lesser—mathematician.

The rejection was a blow to Charles's pride, but there was

Scientist Versus Society

still work to be done and Babbage proceeded to study diligently for the master's degree, which he was awarded in 1817.

Armed with this additional qualification, the young man again applied for a teaching position, this time at the University of Edinburgh. Once again he was refused; the reason given was that he was not a Scot.

For a person so accustomed to success, this second rejection would not soon be forgotten. Yet it was hardly disastrous. Babbage had sufficient means, thanks to his father and to Georgiane's, to support his family in comfort. The couple now had two children, and Charles and his family moved into a comfortable apartment over the carriagehouse attached to Benjamin Babbage's mansion in London.

There was ample space for the family, as well as space for a workshop where Charles could tinker to his heart's content. Moreover, there was time to explore some of the ideas that had so long fascinated him. Foremost among them was that of a machine to calculate the kind of tables he had studied and found wanting in his Cambridge days.

Accuracy was of paramount importance—the accuracy that the machines of Pascal and Leibnitz had lacked. But Babbage also would prevent the mistakes of careless typesetters. *His* machine, when it was completed, would be able to print out the results automatically, once they were obtained.

It was an ambitious project, but Babbage set to work with his usual single-mindedness, and it was not long before he had found at least a partial solution to the problem. He decided to base his "Engine" on the principle of differences, which is known to every mathematician. That principle, simply stated, is the process of multiplying by means of addition.

Babbage later drew up and published a table showing

Charles Babbage

exactly what he had in mind. His table, based on the price of meat, was similar to the following:

Pounds	Price	Difference
1	5d	
2	10d	5d
3	1s 3d	5d
4	1s 8d	5d
5	2s 1d	5d

Thus he showed that there was always the same difference in the price per pound. That difference, of course, was 5 pence. And Babbage—or the butcher—could figure the cost of five pounds of prime beef either by adding together the differences or by multiplying the cost of one pound of meat by the number 5.

Babbage went beyond that simple form, however, drawing up more complicated tables as illustrations. In some of these, the *constant* difference was not apparent in the first calculation but only in the second. A table of squares is an example.

Number	Square	First Difference	Second Difference
1	1		
2	4	3	
3	9	5	2
4	16	7	2
5	25	9	2

In some cases, as Babbage himself pointed out, that constant difference—the number that was always the same—was found only in the third or even the fourth dif-

Scientist Versus Society

ference. But such a fact had no bearing on Babbage's invention. As long as there existed a difference at some point, the machine he had in mind would work.

It was five years before Babbage succeeded in constructing the first simple working model of his "Difference Engine." On June 14, 1822, he was able to demonstrate it at a meeting of the Royal Astronomical Society, which he had helped to found, and of which he was, of course, a member.

The enthusiasm of the others was beyond Charles's greatest hopes. If his peers had found so much to praise in the simple machine he had described to them, surely they would shower him with the honors he coveted, herald him as the greatest genius of his time, when he presented a machine infinitely more complicated, one capable of accomplishing what seemed even greater miracles.

He had already posed the questions that must be answered to make this larger dream come true. He had solved the problems, in theory at least. There was only one thing to prevent Babbage from realizing his ambitions—lack of money. His own income, while ample for himself and his family, could never cover the costs of building the engine he had in mind.

But surely—with the approval of the Royal Astronomical Society, and his own established place in the scientific world—that could be solved. His Majesty's government would undoubtedly advance the needed funds.

The next year, Babbage approached the chancellor of the Exchequer, the official who controlled all government spending, and described to him in glowing terms the ways England would benefit from the use of his machine. Navigation of the seas would become infinitely safer, a most important factor to an island nation that depended on shipping for its very existence. Commercial transactions

Charles Babbage

would be simplified and therefore speeded up. The interests of every business, from banking to building, would be promoted. Indeed, there seemed no end to the advantages that would accrue to England from the new invention. Under those circumstances, it would surely be in order for the British people as a whole, through their taxes, to help pay the cost of developing the engine.

The chancellor was understanding, even sympathetic, and agreed to support the project, which was to be completed in three years. But the agreement between the two men was vague and never written down, a point which would plague their relationship ever after.

Now heady with success, Babbage plunged into his work. But the estimated three years dragged into four, and the engine was still far from finished. Too often, the inventor had scrapped the plans—and the work already accomplished—as a new and better idea struck him.

Then, in 1827, Charles Babbage was suddenly faced with a series of personal tragedies. In quick succession his father, his beloved wife, Georgiane, and two of their eight children died. To add to his ordeal, many of the men who had once supported his work now voiced doubts about it, and those who had been doubtful became actively hostile. One—Sir George Airy, the Lucasian Professor of Mathematics at Cambridge—went so far as to pronounce the whole thing "a humbug." Sir George later became the Astronomer Royal, as well as the most implacable of all Babbage's many foes. Babbage, in turn, let loose his fury on the other man, losing no opportunity to attack or belittle him.

His nerves frayed and his life shatterd, Babbage was sent on a European tour by his physician. In spite of strict orders to the contrary, he continued to supervise the work on his Difference Engine, even when traveling.

Scientist Versus Society

He had at last reached Rome, making his trip by horse-drawn carriage—one he had designed himself in order to enjoy the maximum comfort—when the news reached him that he had been appointed to the Chair of Mathematics at Cambridge University. He was to be the Lucasian Professor there, holding the post that Sir Isaac Newton had once held and that his archenemy, Sir George Airy, had recently vacated. It was a signal triumph for the still young mathematician, a great honor. But it was one he was not inclined to accept.

He was concerned about when he would find time to work on his own invention if he were required to live at Cambridge and lecture at least once a year, as every other Lucasian Professor had. Babbage had actually written a letter declining the post when his friends persuaded him to change his mind. And so he handily worked out a solution to the problems posed. Although he was paid between eighty and ninety pounds a year—a significant sum in those days—he neither lived at the university nor gave a single lecture there!

Babbage was still completely engrossed in the idea of the Difference Engine when he returned to England. But once again he was short of funds to pay the enormous expenses involved. So, as before, he went to the chancellor of the Exchequer, who gave him a grant of £1,500. Moreover, it was agreed that a fireproof vault would be built for his designs and papers, as well as a workshop for his personal use.

But Babbage was soon back, hat in hand, asking for more. Again and again his requests were granted, until a total of £17,000 of the British taxpayers' money had been provided him.

In spite of the generosity of the government, the work proceeded at a snail's pace. Too often, Babbage abandoned half-finished work to begin all over again, basing his new endeavors on a fresh idea. Even without such a propensity, though, the obstacles he faced would have been enormous.

The Difference Engine required countless cogs and wheels, nuts and bolts, cams and links and shafts. Yet there were no standard, interchangeable parts available, as there are now. There were not even tools with which to craft them with the essential precision. It was up to Babbage to have the tools manufactured, by a toolmaker he employed himself, before he could even begin his work.

All these difficulties escaped Babbage's critics, who became more and more impatient as time passed and little progress was made. One article, appearing in a respectable journal, demanded that he account for the money he had received, implying that government funds had been wasted on a dismal failure and that Babbage had concealed the true facts of the matter.

The embattled inventor lashed back in 1830 with a paper, *Reflections on the Decline of Science in England and Some of Its Causes.* In it, he attacked the neglect of science in the universities, as well as the neglect of both science and scientists by the government. He argued that pure science should be not only encouraged but pursued for its own sake. He deplored the dilettantes who dabbled in the field. Above all, he called for sweeping reforms of the Royal Society. The time was past, he insisted, when men who had never published a scientific paper should be elected to that august body merely because of their wealth or social rank. Irritated that such men frequently were appointed to head the group, he demanded free discussion and the democratic election of officers.

Babbage continued to work on the Difference Engine for another three years. But in 1833 a stroke of disastrously bad luck brought the work to a complete halt. When Babbage, waiting for government funds, fell behind in payments to the engineer who was to construct the machine, the employee quit, taking with him all the tools that had been built at the inventor's own expense.

A less obstinate man might have been completely crushed by such a catastrophe. But Babbage already had in mind a new and far more complicated machine, his Analytical Engine. This was nothing less than a mechanical model of the modern digital computer! There would, of course, be no electrical switches, no tubes or relays to perform its near-miraculous work. Yet it would be capable of carrying out any mathematical computation. Like the modern computer, it could compare numbers and make "judgments." It could even work on previously obtained results—that is, it would have what computer specialists now called memory.

The entire operation was based on the use of punched cards, like those used—along with tapes—today. Babbage borrowed the idea from the scheme for weaving patterned fabrics that had recently been developed in France. The inventor, Jacquard, was able to make the most complicated designs—and even to weave a self-portrait—by passing the threads of the warp through a series of cards. The warp threads, which must be moved together, were also fastened to a single rod. When the cards pushed against the threads, the rod was moved, lifting the threads that did not pass through the holes. The shuttles carrying the weft then moved through the threads still in place, to form the pattern.

On the cards Babbage used, he employed the binary system of mathematical notation, as is still done. In this system, instead of the Arabic numbers 1, 2, 3, etc., only the symbols 0 and 1 are used. All the others are translated into this code.

Today, the punched cards of the computer move over electrical switch feelers with almost literally the speed of light. The machine Babbage envisioned and designed, being

Charles Babbage

operated by descending weights and the motion of springs, would have been far slower, taking two or three minutes to arrive at a result. Yet it would have had the same capacity as many modern computers, and many times that of the first computers produced.

In 1834, one year after the defection of his engineer, Babbage again approached the Exchequer, in what proved a vain attempt to raise funds. To support his request, he presented a proposal for the new Analytical Machine. He also asked for an official decision as to what disposition was to be made of the Difference Engine. If a further grant were made, was it to be used for additional work on that, or to begin the construction of his new brainchild?

It was eight years before an answer was forthcoming. And as Babbage had feared, it was negative. The prime minister, Sir Robert Peel, reminded him acidly of the amount already dispensed by the government for this project. Babbage replied, with equal asperity, that he had spent more than that from his own pocket.

Now he placed his hopes in his fellow scientists. Surely they would bring pressure on the government on his behalf. The Royal Society did make such an effort, in spite of the incessant and intemperate attacks Babbage had made on it. But others who had smarted under his stinging remarks proved vindictive. Once again Sir George Airy, now the Astronomer Royal, called the machine "worthless—a humbug." The secretary of the Royal Astronomical Society, the Reverend Richard Sheepshank, complained, "We have got nothing for our £17,000 but Mr. Babbage's grumbling. Those who know how self-opinionated and wrong-headed Mr. Babbage is will have no difficulty in conceiving that his applications to government must have been considered a *bore* by both ministers and secretaries."

Scientist Versus Society

Sheepshank's remark, though, was less cruel than that of Peel, who suggested to the House of Commons that the Difference Engine should be set to calculating the time required before the machine could be put to use.

Embittered and resentful, Babbage at last abandoned the idea of completing his cherished Difference Engine. But his disappointment on that score merely strengthened his resolve to forge ahead with his Analytical Engine. From then on, he was to devote all his efforts and almost all his personal fortune to its development.

The problems Babbage had faced earlier, with his far less complicated machine, had been formidable. But they seemed almost trivial compared to those that lay ahead. It would later take two generations of engineers to solve similar problems in order to build a functioning computer; Babbage, single-handed, undertook the task without a moment's hesitation.

Known techniques were unsuited for drawing his mechanical designs; he developed new ones. No existing algebraic system was capable of describing the moving parts of the machine and their relationships to one another; Babbage invented his own. It was this new algebra that he considered the crowning achievement of his life.

His earlier accomplishments might easily have satisfied a less ambitious or less gifted man. To his lasting credit was the publication, in 1832, of the successful book *The Economy of Manufacturers and Machinery,* in which he explained the method now known as operational research, using it to analyze various manufacturing processes. He made a similar study of the workings of the British Post Office, showing that the cost of collecting, stamping, and delivering a letter was greater than the cost of transporting it. As a result, the British began their "penny post" system by which—as in all modern postal systems—a letter can be sent at a flat rate within a country, regardless of its destination.

Babbage also had published the first reliable "life tables," which were of great importance to insurance companies since they worked out the risks involved and calculated from them the premiums necessary for insurance policies. He became a consulting engineer and worked on the development of railroads. He invented the speedometer and proposed such other improvements as the cowcatcher and broad-gauge tracks.

He found a method of identifying lighthouses by means of occulting—or rhythmically flashing—lights, a system still in use today. He was the first to suggest that the study of the rings of trees could determine cycles of wet and dry weather—an idea that was recently re-discovered and put to use. Babbage was interested in geology and astronomy, and in his last scientific paper, published in 1859, turned to archeology. Much earlier, he had explored the possiblities of such Jules Verne-type dreams as a screw-propeller submarine and had even made preliminary sketches for it. Moving farther into the future—even beyond Verne—he had thought of the use of rockets to boost projectiles.

Nevertheless, it was his Analytical Engine that preoccupied Charles Babbage. He worked on it ceaselessly, and at his own expense, until he had again almost run out of money. Then, in 1848, he again applied to the British government for help. As was now usual, his request was turned down, this time by Benjamin Disraeli, the current chancellor of the Exchequer.

Babbage responded to the refusal with a typical letter —abusive and insulting. He wrote that the machine he proposed to build could "not only calculate the millions the ex-chancellor squandered, but it could deal with the smallest quantities." He had a further word for Disraeli himself, deliberately offending him. "The machine upon which everyone could calculate had little chance of fair play from the man on whom nobody could calculate," he growled.

By then Babbage had alienated most of his friends in the scientific community and almost everyone in the government. He complained bitterly that his own country had ignored his contributions to science, heaping scorn upon him rather than granting him the honors he deserved.

He had set his heart upon a life peerage but had been offered nothing more than a knighthood. Rather than accepting that not inconsiderable distinction graciously, Babbage scoffed at it, stating that it was "fit only for mayors and second-rate men" and that he had no wish "to join the circle of the B-Knighted."

In fact, Babbage had received many tributes from the British people and had been accepted socially by the most prominent of them, including Queen Victoria and Prince Albert. But he preferred to overlook all this and to dwell on the failure of his compatriots to appreciate the two machines that were so dear to him.

It was only on the Continent that the esteem he craved was forthcoming. Babbage had long been in touch with a prominent Italian scientist, Baron Plana, and had told him of his work in a number of letters. Plana had been sufficiently impressed to invite the inventor to a scientific meeting in Turin, held in 1840, at which he would describe the Analytical Machine.

His discourse was received enthusiastically by all present and aroused such favorable interest that Babbage was granted a series of audiences with the king, Charles Albert, with whom he discussed scientific matters. After the third and final meeting, he was presented with a medal that he treasured all

Charles Babbage

his life, as he did the title bestowed upon him of Commander of the Italian Order of St. Maurice and St. Lazarus. Later, when he published his autobiography—which was first refused by his publisher because of his caustic comments on many still-living Englishmen, he dedicated the book to Charles Albert's son, Victor Emmanuel II, commenting that he was the "Sovereign of united Italy, the country of Archimedes and Galileo." It would have been difficult to ignore the fact that Babbage was comparing himself to two of the finest scientific minds in history!

One of those who heard Babbage and was impressed by him was L. F. Menebrea, a mathematical engineer. He understood the Analytical Engine so well, and was so convinced of its practical application, that he wrote a paper on it, describing it clearly. Called *A Sketch of the Analytical Engine Invented by Charles Babbage*, it was published in October 1842 in the *Bibliothèque Universelle de Genève*. Menebrea summarized his remarks with the statement: "Thus the idea . . . is a conception which, being realized, would mark a glorious epoch in the history of the sciences."

Babbage was delighted to learn, not long afterward, that the work was being translated into English by Ada Augusta, the countess of Lovelace, who added her own notes, making it the most lucid exposition of the subject ever to appear—far more comprehensible than any explanation Babbage himself ever gave.

Lady Lovelace, the only legitimate child of the poet, Lord Byron, was a brilliant mathematician. She had become a fast friend of Babbage—one of the few who had confidence in him.

Babbage was now desperately in need of money to carry out his work. He hoped that Lady Lovelace, through her own connections with people of wealth and importance, could help him. But those who might have been of some assistance, either by themselves or through their influence, were thoroughly tired of the irascible genius. "Babbage rhymes with cabbage" was a current witticism, and it was repeated often.

For a while the scientist, in company with Lady Lovelace, tried to raise the required funds by betting on the horses. Together they worked out a system that was mathematically foolproof. In practice, though, it was disastrous.

After that debacle, Babbage toyed with other ideas, equally impractical. He planned to write a three-volume novel and was certain it would bring him a fortune until a writer friend explained the economics of publishing to him. Another idea was to construct a machine that would play tic-tac-toe and to exhibit it at fairs and circuses. Since Babbage planned to embellish it with figures of bleating lambs, crowing cocks, and children joyfully clapping hands, he was certain there would be throngs willing to pay several shillings to see it. But once again he was persuaded to abandon his idea—no one could compete with the enormously popular circus entertainer, General Tom Thumb.

Babbage applied for government jobs, hoping at one time to be appointed master of the Mint. But he had offended so many people in power that no one would put in a good word for him.

And he continued to offend them. He accused Sir Humphrey Davy, the president of the Royal Society, of swindling between £3,000 and £4,000 of the society's funds. As for the society itself, he announced that it was "a collection of men who elect each other to office and then dine together at the expense of the society to praise each other over wine and to give each other medals." The numerous committees that were formed—and on which he was rarely asked to serve—prompted the comment: "Occasionally a few simply honest men are to be found upon a committee. They are useful to give

Charles Babbage

a high moral tone to the cause—but the rest think them bores."

It was not only those in lofty positions whom Babbage attacked. He was equally contemptuous of the lowly. Even his own children were terrified of him. But the bane of his existence were organ grinders and other street musicians.

Time after time he had such offenders hauled into court, complaining that they had become a nuisance. He bombarded the London *Times* with letters about them and sent still others to members of Parliament. He soon became known as an eccentric—a crackpot—and was the laughing stock of much of England.

The letter Babbage sent to Alfred, Lord Tennyson, merely added fuel to the fire, making him appear even more ridiculous. Tennyson had recently published a highly praised poem called "The Vision of Sin." In it he wrote:

> Every minute dies a man,
> Every minute one is born.

Babbage took exception to this, on purely statistical grounds. Were it true, he informed Tennyson, the population of the world would remain the same, when it was quite obvious that it was constantly increasing. He therefore suggested a change. The lines in question should read:

> Every moment dies a man
> And one and a sixteenth is born.

Even such an absurdity was not enough for Babbage, who went on, "I may add that the exact figures are 1.167, but something must, of course, be conceded to the laws of meter."

To everyone's amusement, Babbage also set about to work out the possibility of miracles. The chance of rising from the dead, he calculated, is one time in a trillion.

Scientist Versus Society

As Babbage became more and more paranoid—more and more obsessed with the idea that he had been persecuted all his life—his fellow Londoners began to treat him much as small boys treat the village idiot, following him along the street, making sport of him. It became a custom among the more spirited citizens to end a night on the town by stopping under Babbage's window late at might to serenade him with the blast of horns and the roll of drums.

Babbage grew old almost alone. Only two of his eight children survived him and he had alienated both, although they were reconciled at the end. He clung to his cherished dreams of the Difference Engine and the Analytical Engine, although he knew that he would never see those dreams realized. Once, a short time before he died, he confided to a friend that he could not remember a single happy day in his life. "He spoke," the friend said, "as if he hated mankind in general, Englishmen in particular, and the English government and organ grinders most of all."

When Charles Babbage died, on October 18, 1871, only two friends joined the handful of relatives at his grave in Kensal Green Cemetery in London. The *Times* wrote, maliciously, in his obituary: "He lived to be almost eighty in spite of organ-grinding persecutions."

In the ensuing years, Babbage, if he was remembered at all, was remembered for his crotchety disposition. It was not until 1944 that a British writer, L. J. Comrie, himself an inventor and developer of a calculating machine, gave Babbage his due. Comrie's extravagant praise for Babbage was matched by his harsh words for England. "The black mark earned by the government of the day more than a hundred years ago for its failure to see Charles Babbage's Difference Engine brought to a successful conclusion has still to be wiped out. It is not too

Charles Babbage

much to say that it cost Britain the leading place in the art of mechanical computing."

But it was not only in his conception of the Difference Engine and the Analytical Engine that Charles Babbage was ahead of his time. It was also in his realization that the subsidizing of scientific research was—and must be—the province of modern governments everywhere.

Babbage once said that he would be glad to give up the rest of his life if he could return to earth in five hundred years and explore, in the company of a guide for three days, the scientific discoveries after his death. If he were to return today, a little more than a hundred years after his death, he would see his great dream of a machine to solve algebraic problems come true. Moreover, he would at last receive his long overdue recognition.

For once, Charles Babbage would enjoy the one happy day that had evaded him all his life.

III
ADA AUGUSTA, THE COUNTESS OF LOVELACE

Ada augusta byron was the only legitimate child of George Gordon, Lord Byron, one of England's greatest poets. Both brilliant and beautiful, she was almost as gifted a mathematician as Charles Babbage, whose friend she became. Unlike Babbage, who came from a wealthy and respectable, but nevertheless bourgeois family, Ada Augusta could point to a long line of forebears among the nobility.

Her mother's family traced its lineage back to Henry VIII—and before. Her father's family was even more distinguished. The Byrons traced their lineage as far back as 1066 and the Norman conquest of Britain, while on the maternal side there were connections with Mary Stuart, queen of Scots.

But their more recent history had been far from glorious.

Lord Byron's grandfather had presumably gone mad. His great-uncle, the fifth Lord Byron, had been tried for the murder of a neighbor and subsequently was known as "the wicked lord." Although he had inherited a fortune, he had quickly lost it and left little to his heirs.

Ada Augusta, the Countess of Lovelace

Byron's own father, known as "Mad Jack" Byron, had restored the family fortunes by marrying an heiress, whom he persuaded to leave her husband and elope with him. After her death, at the birth of her daughter Augusta, he had married again. It was a miserably unhappy marriage, marked by constant quarreling. When the couple's only child, George Gordon Byron, was two and a half years old, they separated. A year later John Byron was dead, the new fortune also dissipated. The young boy was brought up in poverty by his shrewish mother, who alternately indulged and berated him. When he was six years old, however, the title of sixth Lord Byron passed to him, along with the family estates.

As befitted a young man of his rank, he was educated at the famous English public school Harrow, and at Cambridge University, before leaving England to travel in the then mysterious lands of the Middle East. He was moody and shy, almost morbidly sensitive about his deformed foot, which detracted from his otherwise great physical beauty.

Lord Byron was only twenty-four when his greatest work, *Childe Harold*, was published. The poem was an immediate sensation. He was back in London by then. Suddenly the world, which had scarcely noticed him before, was at his feet. The most brilliant hostesses sought him out for their receptions; the most beautiful women begged for introductions.

One of the few young women who refused to follow the flock was Anne Isabella Milbanke, nicknamed Annabella, up from the country for her first season in the capital. Annabella was an heiress and therefore more than interesting to the great number of youthful and impoverished men seeking wives—and fortunes—and willing to trade titles for them. She was certainly not beautiful but she was attractive, with a slender figure and an animated manner that found favor with many.

Scientist Versus Society

Like Byron, Annabella was an only child. But she had been the spoiled and petted darling of elderly parents who for many years had longed for children. Their lives revolved around their daughter, and she was given every possible advantage, encouraged in every endeavor.

She had not gone to school—at that time, young women of her class rarely did—but she had been taught to read and write and do her sums by her governess while still a very young child. Later there had been dancing masters, riding masters, and tutors. Annabella learned French, after a fashion, and Latin. She studied history. She wrote poetry—and even sent some of her verse to Byron for his opinion. But her real talent seems to have been in mathematics. Lord Byron later was to call her his "Princess of the Parallelograms," still later, the "Medea of Mathematics."

Annabella was most certainly intelligent. Moreover, she devoted her good, but not brilliant, mind to whatever would improve it. The family library was large, but the young girl avoided anything so light as a novel, to delve deep into the most serious works. She never hesitated to believe in her own superiority or to criticize those she felt were inferior. And they were many.

Even in speaking of her father, at the early age of ten, she could reprove him by stating that he was "both good and bad—like everybody else." It apparently never occurred to her that she, too, was "both good and bad." She was, instead, Anne Isabella Milbanke, a paragon of virtue.

It was this side she presented to the world when she arrived in London. She had been taken under the wing of her aunt, Lady Melbourne, her father's sister and the most fashionable as well as one of the most scandalous figures of English society at the time. Annabella, of course, was invited everywhere, entertained—and courted.

Ada Augusta, the Countess of Lovelace

Her reaction to her social success was most often one of scorn. Balls bored her; she considered the opera a waste of time.

She had a series of suitors and refused them all. When, with the connivance of her aunt, Lady Melbourne, Lord Byron proposed marriage, she refused his offer, too.

But she quickly reconsidered. Not only was Byron the catch of the season, if not the decade—he also was a poor, lost soul. He was a tormented creature, a sinner—Childe Harold himself, the romantic hero, passionate, wild, libertine.

Aware of what she had let slip through her fingers—the prestige of being Lady Byron was in itself enough to stir her to action—she began a long correspondence with the poet. Byron responded and, after almost two years, again asked Annabella to marry him.

This time he was accepted. Dutifully, he set off for the family homestead on his first formal visit. Byron spent a short time there, making the acquaintance of his future father-in-law, whom he liked, and his future mother-in-law, whom he detested. He also spent time with Annabella, properly chaperoned, of course. Under the circumstances, the young couple had little time really to get to know one another. Byron left at last, to go to London to arrange a number of pressing business affairs and to consult his lawyers about a marriage settlement.

He was in no hurry to return. Annabella—and marriage to her—awaited him, and he postponed that as long as possible. When he could procrastinate no more, he went again to Seaham, the Milbanke estate, stopping on the way to visit his half-sister, Augusta, the one person who was truly dear to him.

Byron and Annabella were married at last on January 2, 1815, at Seaham. They left for a honeymoon at another estate

45

Scientist Versus Society

belonging to Annabella's family and spent three weeks there.

It was a disaster.

Byron had not really wanted marriage. Now that the ceremony had been performed, he was more certain of that than ever. Above all, he was certain that he did not want Annabella Milbanke as his bride.

The couple returned to London, stopping on the way to visit Augusta. It was soon evident that Byron much preferred the company of his half-sister to that of his wife.

They went on to London, where Byron alternately ignored and tormented his wife. By now she was pregnant. Her health had never been good—Byron complained on their honeymoon that she was ill two days out of every three—and it became worse as she awaited the birth of their child. And to add to her troubles, her husband was so deeply in debt that bailiffs took up residence in the mansion they had rented, ready to seize their furniture and any other movable possessions.

The bailiffs were there when Ada Augusta, their only child, was born on December 10, 1815. It was just a little more than eleven months after the marriage of her parents; it was also the end of that marriage. As soon as she was well enough to leave, Annabella did, claiming that her very life was in danger. She took her month-old daughter with her. Byron was never to see his child again.

Within a few weeks, Lady Byron—with a battery of lawyers advising her—sought to obtain her husband's agreement to a separation. It was a long and bitter battle, and during this time the baby, Ada, was left with her grandmother, Lady Milbanke.

Because of Lord Byron's prominence, the separation stirred up a tempest. Lady Byron refused to explain her reasons for leaving her husband, which led to a spate of rumors. One charge, which Annabella refused to deny, and which has therefore been generally accepted, was that Lord Byron had been in love with his half-sister, Augusta Leigh, and that he had fathered one of her children, Elizabeth Medora.

There were other charges as well—and again Lady Byron in effect confirmed them by refusing to deny them. The result was that the poet's name was blackened and Annabella won the sympathy of the public.

Inevitably, young Ada was drawn into the controversy. To make certain that her father would not attempt to gain her custody and take her abroad, where he had fled, she was made a ward of the court.

Byron was allowed no contact with his daughter, even through letters, although he often sent gifts to his half-sister Augusta, to be forwarded to Ada. The bitterness of Annabella's family was so great that *her* mother—Ada's grandmother—tried her best to prevent the child from learning anything about her father and went so far as to stipulate in her will that Ada should never even see Byron's portrait until she was twenty-one.

It was impossible to ignore Lady Milbanke's desires. It was she who had inherited a large fortune, one that eventually would go to her daughter and then to Ada. In the meantime, though, there was almost nothing. The fortune Byron realized from the sale of the family estate was to go to his half-sister on his death. During his lifetime, his wife received only £500 a year, an amount clearly insufficient to meet her needs and those of her child.

There was little help, at the time, from Lady Milbanke. England was suffering a general depression, and her own—and her husband's—properties showed a loss rather than an income. Annabella was unhappy under her mother's roof and she began to travel throughout England, taking the tiny girl with her.

Scientist Versus Society

For many years, Annabella's life was devoted to justifying her actions against Lord Byron. As the scandal had gradually subsided, public opinion had changed. Now her husband was considered the wronged partner, and the idea was intolerable to Annabella. She set about writing her memoirs, destined for Ada. She continued to defend herself wherever and whenever she could. And above all, she went about "doing good."

Young Ada often was the unhappy victim of her mother's well-meant intentions. And in spite of the economic difficulties the pair faced, she was brought up properly, as befitted a young girl in her position.

Ada's interests were evident very early in her life. Byron had written, when she was only four, that he hoped she would be taught Italian. Later, asking for news of his daughter from Augusta Leigh, he wrote of his wish that "the gods have made her anything save poetical," adding that it was enough to have "one such fool" in the family.

Ada's interests were, in fact, far from literature. Like Charles Babbage, even as a small child she had been fascinated by all that was mechanical. It was her delight to build ships and boats, something Lord Byron also had found enchanting when young. But her special aptitude was for mathematics, and before she was in her teens she taught herself geometry from a textbook that fell into her hands.

Lady Byron was far less sympathetic to the little girl than was the distant, unseen father, then living in Switzerland. Referring to her daughter, Annabella frequently spoke of Ada as "it." But she was determined to do her duty by the child and to see to her spiritual as well as her intellectual development.

Ada was eight when Lord Byron died in Greece, a country where he felt far more at home than in his native England. Told of his death, Ada was heartbroken.

Ada Augusta, the Countess of Lovelace

Lady Byron at the time was still traveling about England, still trying to justify her actions. Her mother, Lady Milbanke, who had cared for Ada, was now gone, too. A year later, the little girl's grandfather died; there was no one left to care for her but her mother.

Annabella hired a governess for her daughter and, with another woman, an old friend, they set off to tour Europe. When Ada showed an interest and a talent for drawing, a suitable teacher was found. She was encouraged to learn the languages of the countries she visited, to explore whatever aspects of the culture most interested her.

Annabella Byron's newest concern was education, and her views on the subject were extremely liberal, far in advance of her times. Young Ada profited from this, as she must have suffered from the moral righteousness of her mother.

Ada was sixteen years old, endowed with a brilliant mind, her intellect developed well beyond her years, when the governess was dismissed. A tutor, far more knowledgeable, far better equipped to teach a budding genius, was engaged. And Ada made even more rapid progress under the guidance of the benevolent Dr. William King.

But not long after he had been hired, Ada suffered a complete breakdown. For months, she was not allowed to leave her bed. Later, she could only walk with the aid of crutches. No one could diagnose the illness, and it was generally assumed that it had been caused by the strain of such concentrated and constant study.

It was obvious that Dr. King had to go. He was soon replaced by a Miss Lawrence who had new ideas on education, more progressive than those current in England, although they were even then being tried out in certain Swiss schools that Lady Byron had visited.

Annabella was delighted with Miss Lawrence's avant-garde

Scientist Versus Society

ideas. She had long been convinced that all the perversions of Lord Byron, which she had strongly hinted at but never actually voiced, resulted from his years at his public school, Harrow.

Along with lessons meant to improve Ada's mind, Miss Lawrence encouraged her to free herself of the conservative ideas that prevailed among the English upper classes. She deplored the hold of the church and the clergy, and Ada soon adopted her attitude, refusing to accept anything that seemed dogmatic to her. Long before it was fashionable, Ada had become a "free thinker."

Before his death, Byron had written to his wife, "Pray let her be musical, if she has a turn that way." Ada had, and soon began to play the violin. She played the guitar as well, and the harp, and became an accomplished musician. Later, she understood the possibilities of the use of the computer in the composition of music. Such a use is common today, but she was the first to perceive it.

Ada Augusta was not the first brilliant mathematician to be a gifted musician, nor the first brilliant musician to be a gifted mathematician. Wolfgang Amadeus Mozart had shown a remarkable talent for the science, too.

When Ada was eighteen, Lady Byron decided to put aside the girl's moral and mental development to turn her into a social butterfly. She would be presented at court, like the other well-born ladies of her age.

Ada charmed all but took very little interest in the proceedings. It was Lady Byron who chose her ball gowns, who saw to it that her hair was properly dressed, that she appeared to her very best advantage. And it was Lady Byron, too, who worried about her daughter's reception. The scandal about Lord Byron's behavior had not been forgotten, although he was now the object of public sympathy. Still, what would people say about his daughter? And many people had by now turned against Annabella. Ada was her daughter, as well. What would they say about Lady Byron's daughter?

They said nothing. Ada was completely at ease and utterly delightful, captivating all whom she met. One ball followed another and she seemed to enjoy everything. She was enchanted to meet many of the great men of the time and wrote to her mother to report on them.

Ada thrived on the attention lavished on her. Annabella was suddenly concerned about her daughter's frivolity and took her off to visit the sordid and depressing industrial cities in the north of England, where she could see the squalor in which many of her countrymen lived.

Dutifully, the young woman trailed her mother through the dingy, dreary factories, where even young children stood before machines, tending them from dawn until dark. To Annabella, they were a pathetic sight. But Ada was so engrossed by the machines themselves that she scarcely noticed the miserable workers.

Not much later, Ada was allowed to attend a lecture at the Mechanics Institute in London, where a model of the Difference Engine of Charles Babbage was exhibited. She followed the lecture on it attentively. As for the machine itself, its workings and its potential were completely comprehensible to her. A friend, accompanying her, later wrote that the others "in the party gazed at this beautiful instrument with the same sort of expression and feeling that some savages are said to have shown on first seeing a looking glass or hearing a gun."

Lady Byron was not only determined to impress her daughter with the wretchedness of the English working class; she was equally determined to point out to her the senseless extravagances of the aristocracy and the follies of both. Accordingly, Ada was taken to a horserace at Doncaster

Scientist Versus Society

where, it was hoped, she would share her mother's indignation at the "desperate gambling" of both rich and poor, as well as the terrible risks to both horse and rider.

Instead, Ada was fascinated by the spectacle. Nothing seemed more exciting to her—and nothing less difficult than winning. Surely mathematics was the key to success! There were definite laws of probability in mathematics that could be applied to gambling.

For Ada Byron, the notion was to prove a calamity.

But at that point she was still under her mother's watchful eye and was not yet ready to put her mathematical calculations to the test.

With the London season over, Ada went back to her mother's home in the country, where she spent much of her time studying mathematics. She became interested in astronomy as well. Eager to broaden her field of knowledge, she began to write to some of the more distinguished intellectual figures in Britain.

Contact with them was easy because of her mother's wealth and social position—and also because she was the daughter of one of England's most famous poets. To Ada, that contact was vital. There was no other way for her to discuss the subjects dear to her, to learn from those who had mastered them.

Had she been a man . . .

Had she been a man, she would undoubtedly have gone to one of Britain's two great universities. For Lady Byron's prejudice against English schools did not extend to the best and the highest of them. Even if it had, she would have been forced to send her child, if only to avoid the disapproval of her friends.

But Ada was a woman. There were simply no universities open to her, none she could have attended. And had there been, her mother would have been subject to the same disapproval for encouraging a daughter's intellectual development as she would have suffered had she stifled that of a son.

Ada had already met Charles Babbage at the Mechanics Institute. He was soon to receive a letter from her. Another letter was sent to the wife of Dr. William Somerville, Babbage's colleague and a fellow member of the Royal Society.

Mrs. Somerville was to become one of Ada's closest friends. Like Ada, she had been denied a formal education because of her sex. Nor had she had the advantages of a succession of tutors and governesses. But she had managed to acquire a vast knowledge, as well as an enviable reputation, through her independent studies. She succeeded so well in her efforts that after her death a college at Oxford was named for her, while a college at Cambridge accepted the superb library she had assembled.

Ada wrote to her often, setting forth her own viewpoint and asking for her reaction. Their friendship flourished and the young girl—a third the age of Mrs. Somerville—found the encouragement she needed.

But with the new London season, Ada returned to the social scene. There were balls again, supper parties after them and dinners before. There were titled young men to woo her, as they had wooed her mother. And there was one she would accept, when she was just twenty years old.

He was William, Lord King, the eighth in that line. Later, he was to become the first earl of Lovelace.

They were married in 1835, and two years later their first child, a son, was born. Ada, rebelling against her mother's unforgiving attitude, named the boy Byron for her father.

Neither she nor Lord Lovelace seemed particularly interested in either that child or the two who followed. Often they were sent off to be cared for by their grandmother, Lady Byron, as Ada had been sent to *her* grandmother, Lady Milbanke. Just as often, they were simply left with nurses or servants.

Once again, Ada resumed her studies. Now she was in contact with the celebrated Professor Augustus de Morgan, asking constant questions, receiving answers, outlining new theories. At her insistence, Lord Lovelace invited Charles Babbage to spend a few days with the family one Easter.

The two became the best of friends. The elderly mathematician—he was exactly the same age as Ada's mother—was bewitched by the young woman whom he soon called his "lady-fairy." He became a frequent visitor, always welcome, sharing many of Ada's interests although never that in music.

When Ada was twenty-eight, she undertook the task of translating Menebrea's work on Babbage's Analytical Engine. It was to take her two years.

Such an act would have been unthinkable for most women in her position, had they even been capable of it. But Lord Lovelace was extremely proud of his wife's intellectual gifts and permitted her the great freedom of writing and publishing her work. Moreover, he even helped her, by inking in drawings and correcting proofs. Lady Lovelace could not sign the paper with her own name, of course. But it did appear under her initials, A. A. L.

There was speculation throughout the academic world when the paper was published. Who on earth was A. A. L.? It hardly mattered. The work was so impressive that everyone commented on it.

It was far superior to anything written by Babbage himself. Lady Lovelace not only could understand his work but also

Ada Augusta, the Countess of Lovelace

could explain it, making it intelligible even to the layman. And Babbage, for all his mechanical genius, had seemed unable to set forth his ideas with equal clarity. As a consequence, he had written very little about either his Difference Engine or his Analytical Engine. In his later years, he was unable even to speak of them coherently.

Babbage, who had seen the translation by the countess, had encouraged her to add a commentary of her own. It turned out to be much longer than the original by Menebrea. With it, she included some notes on Babbage's work on certain theorems of a Swiss mathematician, Bernouilli. The Englishman, however, had made a number of errors that Ada quickly discovered and corrected. Had another done so, the haughty, arrogant Babbage might have greatly resented it. But, despite his chagrin, the two remained firm and fast friends.

In writing about the Analytical Engine, Ada found a pleasingly poetic way of describing its function. *"It weaves algebraical patterns,"* she wrote, "just as the Jacquard-loom weaves flowers and leaves." It was in this paper that she foresaw the possibilities of the computer in the field of music, where it might "compose elaborate and scientific pieces of music of any degree of complexity and extent . . . if the fundamental relations of sounds in the science of harmony and musical composition were susceptible of adaption to the notation and mechanism of the Engine."

She sounded a note of caution, too. Already there were those who were crediting the engine with the power of thought—a mistake often made in modern times—unaware that a computer can do only what it has been programmed to do and that its success is the result of the skill of the programmer. As Lady Lovelace expressed it, "The Analytical Engine has no pretensions whatever to *originate* anything. It

Scientist Versus Society

can do whatever we *know how to order* it to perform. It can *follow* analysis; but it has no power of *anticipating* any analytical relations or truths."

Striking as Lady Lovelace's paper was, it gave a mere inkling of her real abilities. But it prompted her friend and teacher, Professor de Morgan, to write to Lady Byron about her daughter. He spoke far more frankly to her than he had to his student, to whom he merely commented from time to time that what she had expressed was "very good" or "quite right." He restricted his praise to such phrases, he stated, because he believed he should not encourage his gifted pupil too much—were she to apply herself fully to the study of mathematics and to the development of her intellectual abilities, her health was sure to suffer.

He did not hesitate to add that her talent was such in "grasping strong points" as well as "the real difficulties of first *principles*" that, had she been educated at Cambridge, her chance of being senior wrangler (first in the class), as Babbage had been, would have been very much lowered. Cambridge, it seemed, still did not appreciate new ideas. But Ada's intuition and creativity would have made her "an original mathematical investigator, perhaps of first-rate eminence."

Other women, he continued, had, of course, published papers on mathematics: "They have shown knowledge and power of getting it, but no one has wrestled with difficulties and shown a man's strength in getting over them." And at last de Morgan came to the heart of the matter: "The reason is obvious: the very great tension of mind which [those difficulties] require is beyond the strength of a woman's physical power of application."

De Morgan had summed up the attitude toward women of

Ada Augusta, the Countess of Lovelace

genius—and toward women generally—that was accepted without doubt by his contemporaries. Women were inferior. It was that simple. No matter what their gifts, they could never compete with men on their own terms, could never fully develop their talents, their mental powers, because of their supposed physical frailties. And should Lady Lovelace, whose powers of thought were as great as those of any man, become completely engrossed in the application of those powers, "the struggle between the mind and body will begin."

There was already a great void in Ada's life. Part of it was due to her mother's lack of real affection for her. But a larger part was due to her complete lack of knowledge of her father.

Certainly other children had been left without theirs, too. But they were told about them, heard them described, and in a sense grew to know them. They became close to a memory of the fathers they had lost and felt a bond, a sense of security.

But everything about Lord Byron and his life had been deliberately concealed from Ada. For a long time, it had been forbidden to mention his name in her presence. She was nearly twenty before she saw a picture of him. And it was only much later that she was to learn of his great love for her, his tenderness toward her.

She had been moved almost to tears by the news, finding her own love for him awakened, while she had turned in resentment against her mother. Eagerly, she sought to learn more of him, of his works, of his life. Nothing seems to have touched her more deeply, though, than her visit to Newstead Abbey, which for many generations had been the home of the Byrons, and where the poet had spent part of his childhood as well as the years immediately preceding his marriage.

Ada insisted on sleeping in his bedroom, on visiting the part of the stately home that had been most closely associated with him. She spent hours examining the swords and pistols that her father had collected, or merely staring from the window at the garden, as Lord Byron must have done.

At times, Lady Lovelace seemed to identify with the great romantic heroines of her father's poems. Once, in that mood, she appeared at the Queen's Ball dressed in a semi-oriental costume, her long black hair plaited with pearls.

She took an interest in Medora, who was reputed to be not only her cousin but her half-sister, the child of her aunt, Augusta Leigh. Ada had learned, long before Annabella hesitatingly spoke of the rumors concerning Augusta and Lord Byron, what their relationship had been. She had been neither shocked nor surprised and had shown little inclination to share her mother's bitterness. And she showed great kindness and affection for Medora's own young daughter, Marie.

Curiously, Ada took far less interest in her own children, always remaining aloof from them. Her older son, Byron, rebelled and eventually ran away to sea, where he contracted consumption and died at an early age. Her younger son, Ralph, was sent to live with his grandmother when he was just nine, returning home only for an occasional visit. While a daughter continued to live with Lord and Lady Lovelace, she was rarely under her mother's care.

Frustrated in a career, devoid of maternal instincts, and caring little for the usual social life of her class, although she enjoyed the theater and concerts, Ada Augusta sought excitement elsewhere. She found it in horseracing, an interest her mother had so unintentionally aroused in her many years before.

Now it became a passion, one she shared with her husband and, to a certain extent, Charles Babbage. Gambling, of course, was the major part of the attraction. With Babbage's aid—and with their mutual knowledge of mathematics—they devised a system that was expected to win huge fortunes for all.

Mathematical rules can be applied to many games of chance, and Babbage had even written a paper on the subject. He explored the mathematical probabilities of certain cards turning up with predictable regularity. There is a regularity, too, in the throw of dice or the turn of a roulette wheel. But the scheme of Lady Lovelace was doomed to fail; chance has little to do with sports.

No matter what the laws of probability might predict, too many other factors enter in. The condition of the track, the spirit of the horse, the health of the jockey—each contributes to the outcome. And Lady Lovelace failed to take this into account.

She and the others began to lose heavily. Eventually both Lord Lovelace and Charles Babbage abandoned the idea of a winning system and simply stopped betting. But Lady Lovelace went on and on, certain that the next modification in her calculations would change her luck and produce a winning streak.

It was not long before she found herself deeply in debt and forced to borrow from the most unscrupulous moneylenders. She was soon threatened with ruin. In a last, desperate effort, she pawned the family jewels, which were eventually redeemed by Lady Byron.

Ada was physically ill now and retreating more and more into a world of her own. She became something of a mystic, convinced that she possessed supernatural powers. Once, writing to Babbage, she compared herself to the devil. On another occasion, she wrote that she would accomplish that "which no purely mortal lips or brains could do."

But for all her belief in her own unearthly powers, she

Scientist Versus Society

proved as frail and foredoomed as every other creature. She was only thirty-six when it was discovered that she suffered from cancer. In the following months, she endured excruciating pain.

During her illness, she and her mother were at last reconciled and Lady Byron was at her daughter's side when she died on November 27, 1852. She had asked to be buried beside her father, who also had died at the age of thirty-six, and her wishes were respected.

She was mourned by family and friends, who keenly felt the loss of a witty and charming woman. But it was not until Babbage's work had come to be appreciated that Ada Augusta, the countess of Lovelace, was recognized for what she was: a mathematical genius who was prevented from developing her unique ability because she was a woman.

IV
GREGOR MENDEL

It was hardly to be envied—that life of the peasant into which Johann (later Gregor) Mendel was born. The work of tilling the soil was back-breaking. The hours in the field were long, beginning with the rising of the sun and ending with its setting. And the rewards were pathetically small.

With a good harvest there would be enough to feed a growing family, but the food would be plain in the extreme. Black bread and cheese were staples, along with potatoes and such root vegetables as turnips and onions in the winter, with dried peas and beans to supplement them. If the farm boasted cows, there might be fresh butter; if hens, fresh eggs. But these were luxuries, reserved for special occasions. The peasant's daily fare was nourishing but hardly palatable.

There would be wood, laboriously gathered during the summer months—or more probably there would be lime in that particular region of Moravia—to be burned during the bitter winter.

Scientist Versus Society

There was little more. Like the animals they bred and raised, the peasants of Central Europe in the early nineteenth century had to content themselves with the barest necessities.

But men, women, and children had certain obligations that must be fulfilled. Three days' work must be given each week to the lord of the manor; in return, he was obliged to protect his subjects from menacing intruders, whether animal or human.

The social system in Moravia, then a part of Austrian Silesia but now belonging to Czechoslavakia, was hardly more advanced than what Western Europe had known in its days of feudalism. Perhaps it even lagged behind. In spite of the protection offered by Count Waldberg, Gregor Mendel's father, Anton, had been conscripted and forced to fight in the wars against Napoleon. It had been eight long years before he returned home.

He had at once settled down to the only work for which he was fitted, agriculture. He was fortunate in having a natural talent—a green thumb—and his small farm prospered. He was fortunate, too, in owning the small plot of land that he worked so industriously. Many of his neighbors were still little more than serfs, cultivating land that belonged to Count Waldberg.

And he was fortunate—although he could not have known it then—that Count Waldberg had established a school in the tiny village of Heinzendorf some twenty years earlier. It was there that his son, Johann, would learn to read, write, and do his simple sums.

The son, the second child of Anton and Rosine Mendel and their only boy, was born on July 22, 1822. The birth of a male child is always a great occasion in such a community, and that of Johann was no exception. The proud parents celebrated with a few neighbors, doubtless over a few glasses of wine or

Gregor Mendel

the local plum brandy. Later, of course, the child was properly christened in the village church that served the spiritual needs of the local inhabitants.

Johann was a sturdy child—all his life his peasant origin would be obvious in his face and figure—and blessed with a serene and happy disposition. He was intelligent, too, quick to learn all that his parents had to impart. And later at the village school the monks who served as teachers were delighted with a pupil so alert, one who so readily mastered his subjects.

Those subjects had been expanded, only a few years before, at the insistence of Countess Waldberg, who had made the small school her special interest. Now Johann, in company with the other eighty or so students, was instructed in both natural history and the natural sciences.

The addition of these new courses to the curriculum was violently criticized, not only in the town but throughout the region. Even the high schools of Moravia had no such classes, and officials of the church, which supervised all education in that Catholic country, sent a special inspector to investigate "this scandal." But the countess staunchly held her ground. Not only would the boys under her patronage continue the studies she had initiated but they would be taught the principles of growing fruit and keeping bees as well.

The school in Heinzendorf was an elementary one, for children in the lowest grades. But the town of Leipnik, about thirteen miles away, boasted an intermediate school. Two of Johann's schoolmates—exceptionally bright boys like Johann—were now attending classes there. It was soon obvious that the young Mendel belonged with that select group.

Accordingly, Johann's teachers approached Anton Mendel, proposing that he permit his son to continue his education in Leipnik. The father listened, carefully considering the idea.

Scientist Versus Society

He was proud of the boy and as anxious as any fond parent to give him every possible opportunity to enable him to get on in the world.

But Anton was plagued by his own impoverished condition. He had recently torn down the family's ramshackle house and erected a new one, a house still modest but almost luxurious by comparison. That had taken all the cash he could spare. The little else he had carefully hoarded must go to improving the farm itself. Further schooling for Johann seemed far beyond his means. Moreover, Johann was needed at home, where his strong and willing hands would speed the spring planting, the autumn harvest. Again and again, Anton Mendel turned the idea over in his mind or thrashed out the pros and cons in discussions with his wife. At last, with considerable reluctance, he agreed. Johann was to go to Leipnik, where he would be trained for "the arts, sciences, and commerce."

The eleven-year-old boy applied himself diligently to his work, aware of the enormous sacrifices of his parents and grateful for the opportunity they had provided him. At the end of a year, he was ready to enter the high school in Troppau, even farther away from home.

Once again, Anton Mendel allowed the boy to do as he wished. But, in spite of his sincere desire to help his son, he could not possibly stretch his pitifully small income to include the fees for board and tuition that must now be met.

The lack of money seemed an insurmountable barrier to any further education for Johann. But at last an arrangement was worked out that satisfied both school authorities and Johann's parents. It was decided that Johann would pay only part of the normal charge for board at the school. However, being on "half rations," the number of meals he was allowed to eat there would be proportionately restricted. To sup-

Gregor Mendel

plement Johann's meager diet, Anton and his wife would send him supplies of bread and butter from the farm at Heinzendorf whenever a horse and cart set out to make the twenty-mile journey.

The period Johann spent at Troppau—he completed six years of schooling there—was harsh beyond belief. Almost always overworked and frequently underfed, the boy struggled constantly against overwhelming odds.

There was no respite for him in the summer, either, when vacation came. Those days were spent on the farm, where his help was needed in the fields.

Then, in 1838, in his fifth year at the high school, there came another setback. Word was brought to the youth that his father had been seriously injured in an accident at work, when a falling tree had crushed his chest. Hopelessly crippled, Anton Mendel could barely manage to keep up the farm. Any further payment of school fees—even reduced ones—was out of the question.

The only way that Johann could continue his education was to pay for it himself. And there was only one way to earn the money he so desperately needed—by tutoring less gifted youngsters.

Mendel was able to enroll in a special course to qualify for such a position. But the strain of the additional work, coupled with constant undernourishment, was too much for him and he became so ill that he was forced to drop out of school. He returned to his father's farm, staying there for months. That fall, however, he was back at Troppau for his final year at the high school. And in August 1840 he graduated. His grades had been good—a remarkable achievement considering the obstacles in his way.

But now Johann Mendel clutched his coveted and hard-earned diploma and pondered his future. He could scarcely

Scientist Versus Society

return to the farm and to manual labor—or to his father's way of life—now that he had been educated far above the station of a peasant. Yet he could hardly embark on a career without further study. There was always the church, the one institution that unfailingly offered opportunities to young men of brilliant minds, humble origins, and no means at all, to whom other callings were closed. But Mendel was not yet ready to fall back on this last recourse. Instead, he would try to make his way himself, as he had already been forced to do. And he would begin by entering the Philosophical Institute in the city then called Olmutz but now known as Olomouc.

Once again, Mendel counted on supporting himself as a tutor. He had had hard sledding before, but now, with six years of high school behind him, he was certain that things would take a turn for the better.

Nothing could have been farther from the truth, as the aspiring young man soon found out. Had he had friends in Olmutz, they might have steered floundering students his way. But Mendel was alone, lacking the connections many other students—his competitors—had. Nevertheless, he somehow managed to finish his first year at the Philosophical Institute, part of the university in Olmutz.

Mendel's second year was, if anything, more difficult than his first. From time to time he was forced to drop out because of illness, which he attributed to his constant anxiety about money. And word from home was discouraging. Anton Mendel's health had become so bad that he was forced to sell his farm.

The buyer was his own son-in-law, the husband of his elder daughter, Veronika. Although only a trifling sum was involved in the transaction, Anton did his best to provide from it some measure of security for his son and his younger daughter, as well as for himself.

The elder Mendel was to receive some cash outright as well as a small pension for life. Johann was to receive a far lesser sum each year that he continued his studies. Should he, too, become incapacitated, he was to be granted free living quarters on the farm, as well as a tiny plot of land to cultivate although he would never own it. As for the younger daughter, Theresia, her brother-in-law was to provide her with a dowry—the money she would settle on her husband when she married.

In spite of the promise of that money from home, it seemed impossible that Mendel could go on. The amount he was to receive was too small—far less than he required for even the marginal existence to which he was accustomed.

It was then that Theresia came to her brother's aid. She would be willing, she said, to renounce her right to a dowry if Veronika's husband would increase the allowance to Johann. It was a selfless act; the young girl knew only too well how slim would be her chance of finding a husband if she were penniless.

Theresia's generosity—which Mendel later was to repay many times over—enabled him to return to Olmutz and finish his second year. Moreover, his own fortunes took a turn for the better and he found a few pupils who could pay enough to supplement his limited income.

At the end of that year, 1842, when he was twenty, Johann Mendel again took stock of himself. In spite of Theresia's magnanimous gesture, his funds again were alarmingly low. And although years of hard work and deprivation had brought him an education, there were huge gaps in it, vast areas of knowledge that he had not even begun to explore. Above all, there was the fact of his background. He was a peasant's son, and that in the highly structured society of Central Europe, where opportunities were almost invariably denied to all but the wellborn. Mendel summed up his position many years later in a brief autobiography in which he wrote of himself in the third person: "It had become impossible to continue such strenuous exertions. It was incumbent upon him to enter a profession in which he would be spared perpetual anxiety about a means of livelihood. His private circumstances determined his choice of profession."

The only profession open to him was to become a man of the church. He had long before considered the priesthood. Now he made his decision and applied for admission to the Koniginkloster, the Augustinian monastery at Altbrünn, today Brno, Czechoslavakia, but then the capital of Austrian Moravia.

He was well recommended and his application was accepted. On September 7, 1843, Johann Mendel entered the order as a novice, taking the name of Father Gregor, the name by which he would be known to posterity.

The monasteries of Eastern and Central Europe still occupied the place they had held in Western Europe in the Middle Ages. As barbarian hordes had overrun once civilized lands, the monks had retreated to their cloisters to preserve the ideas of the past, the remnants of the disappearing cultures. They had laboriously copied out by hand the books of those earlier times, had recorded history, had preserved the ancient Latin tongue. They had become teachers, too, handing down to the future the knowledge of the past.

The Koniginkloster—The Empress's Cloister—carried on this scholastic tradition. The monastery was the center of the intellectual life of the entire region, its focus of culture. The Augustinians' library of twenty thousand books, a vast collection for the time, was regarded with awe and wonder by the average citizen and with deep respect by the highly educated. Moreover, almost all the monks in residence engaged in independent scientific or artistic work.

But a novice like Gregor Mendel was still required to spend much of his time pursuing his classical studies and participating in the offices of the church. There were periods of prayer each day, as well as periods set aside for contemplation. But, to Mendel's delight, there was still time to work in the monastery garden.

It had been cultivated earlier by another Augustinian monk, Father Aurelius Thaler, who was widely known as a botanist. But he had died only months before Gregor entered the order, so the young man was deprived of the chance to work with a true expert. For the most part, Mendel was left to his own resources and ingenuity. He accomplished much, although, as he later wrote, he "had no oral guidance." But he had the same keen interest he had had as a child, and that interest now served him well.

The struggle for mere survival that had so sapped Mendel's strength in his earlier days was now a thing of the past. The relaxed atmosphere and the very real security afforded by the Koniginkloster put an end to the anxieties and fears that once haunted him. Relieved of such burdens, he could devote full attention to his studies, soon catching up in areas where he had fallen behind while forging ahead in those in which he already had a solid background. Still, it was two years before he was sent by the prelate of the monastery to begin his studies at the Brünn Theological College, the next step in his training. And two years later, in 1847, because of a shortage of those qualified to conduct services, Mendel—who had taken his vows but was still a student—was ordained a priest.

The young man, now twenty-five years of age, had been trained as a parish priest, but he found his new duties so unpleasant as to be almost unbearable. German had been his native tongue. Now he was required to speak a Serbo-Croatian dialect and mere communication with his flock became a problem. Moreover, he could not bear the sight of suffering; visiting the sick, comforting the dying, and consoling the bereaved made him physically ill.

Since Father Gregor clearly was unsuited for such work, the prelate, Father Cyrill, searched for something more congenial for him to do. He soon hit upon the idea of teaching and, although Mendel had not passed the examination that normally was required, it was arranged that he would take over a few classes in a nearby elementary school.

The work pleased the young priest, and the director of the school was satisfied with his new teacher. But Mendel could remain in the school only as a substitute until he had earned a teaching diploma. By now he was anxious for a permanent post, one that would enable him to earn his own living and be independent, like many monks in his order. Accordingly, he prepared to take the examination that would lead to the necessary certificate.

But Mendel's background was sadly deficient in the field that most interested him, the natural sciences. He had had little formal training in the subject—his studies in the monastery and at the Theological College had been restricted to ecclesiastical history, moral theology, and such languages as Hebrew and Syriac—and most of the books at his disposition for independent research were out of date. With no knowledge of the new developments, Gregor Mendel failed miserably. One of his examiners reported that the priest "lacked insight and his knowledge was without the requisite clarity," although he added that the young man "had studied diligently." Another examiner was less harsh in his opinion and was enough impressed by the results—considering

Mendel's lack of formal education—to suggest that he take a further examination that was to be given in Vienna.

If Father Gregor had been ill equipped to answer the questions put to him in Altbrünn, in Vienna he was hopelessly beyond his depth. He managed only a partial response to the problem set him in physics; when it came to natural history—he was asked to classify the mammals and explain their economic value—it was obvious that he lacked even the most elementary knowledge.

In spite of his disastrous showing, Mendel managed to impress one member of the judging panel. When Father Cyrill asked this man the cause of his charge's failure, he answered truthfully that Father Gregor had nowhere near the education needed to qualify as a teacher. However, he added that the young man was both intelligent and industrious, and urged that he be given an opportunity to overcome his disadvantages. As a result, the prelate conferred with his bishop and the two decided that Gregor should be sent to the University of Vienna.

It was 1851 when Mendel at last entered that institution. Two years later, when he left, he had acquired a sound knowledge of zoology, botany, physics, and paleontology. He had learned enough mathematics to solve physics problems and had picked up such practical techniques as using microscopes and setting up physics experiments. But he still had not won his coveted teaching certificate.

Gregor had thoroughly enjoyed his stay in Vienna. There he could speak German again, and so felt completely at home. And he had made firm friends among men who shared his passion for the natural sciences. They were so interested in his ideas that he was proposed for a life membership in the Zoological and Botanical Society of Vienna. He was duly admitted and read a paper—on the devastation of the garden

Scientist Versus Society

radish by a species of small butterfly—at the first meeting he attended, in 1853.

Shortly after, he returned to Brünn to take up his teaching duties at the newly established Brünn Modern School. Earlier, Father Gregor had taught the elementary grades; now his work was in the equivalent of a high school. It was a school that emphasized science to a far greater degree than was common at the time, and when the school suddenly found itself in need of a teacher of physics and natural history, Mendel was asked to take those classes.

He was still a substitute, though, and must remain one until he obtained the necessary credentials. But now there was ample time for study and once again Mendel sat up late, poring over his books. He was able to take on a few private students, too, tutoring them for the very examination for which he was preparing. He even managed some original research and wrote to his friends in Vienna with such enthusiasm about his discoveries that at least one of his letters—describing the pee-weevil, which was damaging crops in Brünn—was read at a meeting of the Botanical Society.

Those two years passed quickly, and then Gregor Mendel packed his small, shabby bag for his final trip to Vienna, his final encounter with his impassive examiners. He faced them on May 5, 1856, confident now, sure of their approval.

At first, all went well. Mendel answered the questions put to him both quickly and correctly, and was rewarded with friendly smiles and nods of approbation. But then a question was asked on a botanical subject.

Father Gregor reflected for a moment, mulling over the answer in his mind. He began to speak at last, clearly and distinctly, adding details, elaborating.

As he continued, he glanced at the faces of those who were to judge him. He noticed the slight frown that furrowed the brow of one man, the tightening of another's mouth. Yet Father Gregor went on until he was interrupted by a brusque question: Did he *really* mean to say that which the learned gentlemen had just heard? Had he not made some mistake? Was that, truly, the answer he wished to give?

It was, Father Gregor insisted. It was indeed.

The panel pursued the matter, making it plain that the young man's answer was unorthodox. It would be better, the judges suggested, if he were to retract it and to respond with something more to the liking of his listeners.

But Mendel refused, insisting stubbornly on the veracity of his viewpoint. Again and again, he was prodded to change the substance of what he had said, to state what the others wished to hear. But Father Gregor refused to give in, to parrot words that, although acceptable to the rest, he considered false.

The ordeal was ended at last, the examination finished. Once again, Gregor Mendel had failed it.

He returned to the monastery in Brünn, saddened but not embittered, refusing thereafter to speak of what had happened. Again he plunged into his teaching, aware that the permanent post of which he had dreamed so long was forever beyond his grasp. But he was consoled by the esteem of his students and the admiration his colleagues showed for his work.

The argument in Vienna had by no means dampened Mendel's interest in scientific research. If anything, this served to increase it. He had been challenged. And how did one answer such a challenge if not by continuing to search out the truth?

Sequestered though he might be in the Koniginkloster in Altbrünn, Mendel was nevertheless far more in touch with the outside world than he had ever been before. His studies were no longer confined to those that would—or should—lead to

Scientist Versus Society

the much-desired diploma. Now he was free to range about at will, sampling whatever new ideas presented themselves.

There was much to stimulate him intellectually at the regular sessions of the two scientific groups that met in Brünn and of which he was a member—the Werner Society, busily engaged in collecting sufficient information to publish a geological map of Silesia and Moravia, and the Agricultural Society of Moravia and Silesia. Later he would help to found the new Brünn Society for the Study of Natural Science. He had friends at the Brünn Modern School whose ideas intrigued him. Above all, he had the time to read the latest scientific books, and at long last the money to buy them.

Foremost among these were the works of Charles Darwin, which had just been translated into German and were now creating a furor there and in the rest of the world. Enthralled by the subject of evolution, and especially the problems of heredity, Mendel avidly perused the texts, making copious notes in the margins and on the endpapers. Not content with what he found in Charles Darwin's works, he delved into others, including that of Erasmus Darwin, Charles's grandfather. Many of the books Father Gregor read so eagerly were actually on the Index—that is, they were banned by the church. That fact never deterred him; he thirsted after knowledge and sought it wherever it was available.

Unlike some of his contemporaries, Mendel was quite willing to accept the theory of evolution as expounded by Darwin. Who could quarrel with the idea that all species of plants and animals developed from earlier forms? And that that development was due to the hereditary transmission of slight variations from one generation to another? Who could deny that the forms that survived were the ones best adapted to the environment?

Certainly not Father Gregor Mendel.

Gregor Mendel

But there was something lacking in Darwin's theory, as even Mendel had to admit. The two vital questions of how and why had not yet been answered. One could accept Darwin's theory of evolution, but someone would have to explain its workings. How could one account for the extraordinary number of *variants* in living things? And, even more important, why did certain descendants of the original plants or animals inherit the new characteristics while other descendants of the very same plants or animals remained unchanged?

Such problems had fascinated Mendel even before he had chanced upon Darwin's work. In an attempt to solve them, he had begun to experiment with mice, which he kept in his small, cell-like room at the Koniginkloster. Crossbreeding white and gray mice, he noted the bizarre results: Some of the young would be gray, like one parent, while some would be white like the other. But there also would be some that were a strange mixture, apparently a new and different breed. Why? Mendel asked himself again and again. How had such a thing come about?

Mendel was not to learn that from his current experiments; he was, after all, still a servant of the church and under its protection. And many of his superiors distrusted both science and scientific research. The bishop himself was protesting that experiments involving the breeding of animals were immoral. So Mendel turned to plants.

He began his research by crossing flowers in order "to produce new color varieties," as he first noted. He added, "The remarkable regularity with which the same hybrid forms continually recurred when the fertilization took place between like species led me to undertake further experiments designed to follow up the development of the hybrids in their offspring." Those further experiments were to take eight years.

Scientist Versus Society

The monastery had at its disposal vast farmlands that surrounded it. But those fields were cultivated, the sale of their produce bringing a handsome income to the Augustinians. The prelate was not about to relinquish either part of the land or part of the revenue merely to indulge what he considered Father Gregor's foolishness. And so Mendel had to content himself with a tiny plot—120 feet by 20 feet—just behind the library. There he planted the different varieties of peas that he had chosen to study.

Peas were simple to breed and to grow, and their fertilization was easy to control, due to the shape of their blossoms. Above all, their characteristics remained constant. Mendel selected seven in all; among them were the height of the plant, the color of the pod and of the pea itself (yellow or green), and the surface of the skin (smooth or wrinkled).

For a valid experiment, Father Gregor knew that he must have plants that bred true. He devoted two years to finding just such varieties. At the end of that time, he had eliminated twelve of the thirty-four with which he had begun. With the remaining twenty-two, he began his work in earnest.

Each spring he planted the seeds, each variety in a separate and distinct plot, each marked precisely with the information he would later need. As the first tiny shoots appeared he watched over them anxiously, watering them when there was little rain, shielding them when the sun was too hot.

When the blossoms appeared at last, Mendel began the laborious process of crossbreeding that was so vital to his experiment. Using a tiny brush, he transferred the pollen of the plants of one variety to the plants of a second. That accomplished, he wrapped the flower in paper or cloth to fend off any bee or insect that might happen by and could fertilize the plant.

Gregor Mendel

In the fall, Father Gregor gathered the pea pods that now clung to the vines. And throughout the winter he sorted them, noting the traits of each and every pod, of each and every pea within it.

Ever since the sexuality of plants had been discovered by the German botanist and physician Camerarius in 1691, others had experimented with crossbreeding. Many had even followed much the same procedure as Mendel. But none of the others had dealt with generation after generation; none had used control plants as modern scientists do. And none had introduced statistical methods to biological research. Mendel alone had foreseen both the difficulties of the task he had set for himself and the means to achieve the desired results.

He was prepared from the beginning for setbacks, for wasted efforts, for false starts. But at first the results of Mendel's experiments were as encouraging as had been those of his predecessors. When he crossed the pure yellow peas with the pure green peas, he found that the offspring, which he called hybrids, were inevitably yellow. With an enthusiasm that would later prove unwarranted, he concluded that yellow, which he called the dominant characteristic, would always prevail over the green—the recessive characteristic.

Almost certain of what the results would be, Mendel went on to cross his hybrids with other hybrids, expecting the dominant characteristic to appear in each successive generation.

But that is not what happened. In his second generation of hybrids, both green and yellow peas appeared in a confused, even chaotic, jumble. The peas seemed to be mixed up without rhyme or reason, some pods containing only green peas, some only yellow, while some actually contained both!

Perplexed, Mendel set to work again, refusing to abandon

Scientist Versus Society

his venture or to start it anew. His strong religious faith, strengthened by his role as a priest, sustained him in the belief that there must be a pattern to God's world, some method in what was otherwise a muddle. And since that must be so, Father Gregor would find it.

Again and again Mendel crossed his hybrid plants, nurturing them carefully, noting the variations that appeared, the constants that remained. And gradually the scheme that he so diligently sought emerged. If two *hybrid* yellow peas were crossed, three of four progeny would be yellow while one would be green. Therefore, he at last concluded, the ratio of the dominant trait to the recessive in the second generation was 3:1.

Father Gregor discovered another important ratio in the third generation of plants—the generation bred from the four offspring of the original crossing. The green peas of the second generation would produce green peas—that is, they would breed pure. One of the yellow peas also would produce only yellow—or *pure*—offspring. But two of the yellow peas would breed the same way as the first generation of hybrid plants. Like them, they would produce peas with the dominant trait of yellow in the ratio of 3:1. The second ratio that Mendel discovered was therefore 1:2:1—one purebred with the dominant trait, two hybrid peas, and one purebred with the recessive trait in the third generation.

Mendel also found that the same ratios held true in the crossbreeding for *each* of the characteristics he had selected to test. When tall plants were crossed with dwarf plants, the first set of offspring would be tall. Therefore, height was the dominant characteristic. In the next generation, three of the plants—the offspring of hybrids—would be tall while one would be short. The ratio was again 3:1. And once more, in the following generation, Mendel found the ratio of 1:2:1.

But Mendel also learned, to his great surprise, that

Gregor Mendel

characteristics could be inherited independently of one another. While a yellow hybrid pea crossed with another yellow hybrid pea would result in one pure yellow pea, one green pea, and two hybrids, and a tall hybrid crossed with another tall hybrid would result in two tall hybrid plants as well as one pure tall plant and one pure dwarf, such a crossing might easily result in a plant with one of the dominant characteristics coupled with one of the recessive characteristics. A tall plant could have green or yellow peas. Therefore, there were innumerable possible combinations in nature.

It had taken Mendel two years to find suitable seeds with which to begin his extensive testing. Another three had passed as he cultivated his plants. It would take three more years for him to tabulate the results and write them up in a brief paper.

In that paper, Father Gregor formulated what are now known as Mendel's three laws. First, he stated that each inherited characteristic was inherited independent of all other hereditary characteristics. Second, Mendel stated that such unit characteristics are paired in living cells but are split—or segregated—in reproduction, so that each new cell receives only one of the pair.

Third, Mendel stated the law of dominance. In every individual there is a pair of determining factors for each unit, one from each parent. When these factors are different, one (the dominant) appears in the offspring, while the recessive character is latent. This latent factor can and does appear in crossbred generations, in a definite and mathematically exact proportion, according to the total number of offspring.

It was 1865 before the Augustinian monk had completed his experiments. But he was content with the results, and more than content to read his little paper at a meeting of the Brünn Society for the Study of Natural Science.

He was somewhat worried, though, about the length of his

report. Brief as it was, it was still too long to be presented at one meeting. But to Gregor's delight, he was offered the chance to speak at a second meeting just one month after the first.

His fellow members listened attentively as Mendel read, and they applauded politely at the end of the presentation.

Mendel waited for questions about his paper, but none were forthcoming and so he nodded at his audience and took his seat. Certainly at the next meeting someone would have some comment to make, some query.

But the second meeting was much like the first. Again the others listened politely, seeming to indulge the substitute teacher who had failed his qualifying examinations, the parish priest who had failed in that vocation, much as the prelate had indulged him by allotting him the tiny patch of ground on which to grow his peas. As before, there were no questions, no comments—and only the perfunctory round of applause.

Certainly there were some in the audience who had no understanding of what Mendel had attempted and succeeded in doing. His results seemed to them nothing more than a series of useless statistics. But the few who did recognize Father Gregor's remarkable study for what it was were not willing to accept his premises.

That handful included an advanced group who prided themselves on their knowledge of the new Darwinian theory of evolution. Natural selection, as they understood it, simply meant that new species appeared, willy-nilly, from time to time. They did not like to see an uneducated monk challenging that idea.

Moreover, the idea of the survival of the fittest appealed to them. They were—at least by local standards—well-off financially, men with certain achievements to their credit. They enjoyed believing that their successes were due to some

Gregor Mendel

natural superiority that had given them an advantage over others. It was not just in England, which was in the throes of its industrial revolution, that Darwin's phrase was seized upon as an excuse for excesses in the world of business and commerce that violated all humanitarian principles. Such a phenomenon could also be observed in rural communities.

In spite of the lack of enthusiasm with which it was greeted, Mendel's paper was printed in the regular minutes of the society, like all papers read before its members. And the minutes were, as usual, sent out to the 120 university libraries with which the Brünn group was in contact. But like the minutes of meetings before and after, these minutes merely gathered dust on the library shelves.

Mendel made one further attempt to make known the results of his research. The following year, 1866, he wrote to a renowned Swiss professor of botany, Karl Wilhelm von Nägeli, who was then teaching in Munich, explaining his study. Nägeli's great specialty was heredity, and to Mendel it seemed only logical that he, at least, would understand the importance of the work.

But Nägeli had no more comprehension than Mendel's associates in Brünn. He did reply to Father Gregor, suggesting that some of the pea seeds be sent to him so the results could be checked. He also suggested that the monk make the same sort of study on another variety of plant. The one the Swiss had in mind was the genus *Hieracum*, the tawny hawkweed. He could hardly have made a worse choice.

The blossoms of the hawkweed are too tiny to be fertilized artificially with ease; moreover, some of the hybrids are sterile while others are able to reproduce without pollination. Nevertheless, Mendel took Nägeli's advice and proceeded to work conscientiously at his new task, often discouraged but always determined.

Scientist Versus Society

In 1868, however, Mendel's life took a new turn. The aged prelate of the Koniginkloster at Altbrünn had died and a new one was to be appointed. To his great surprise, Mendel was chosen.

The new—and to him, exalted—position brought new responsibilities but also commensurate benefits. At long last, Johann Gregor Mendel was able to repay in kind the generosity of his sister Theresia so many years before, when she had relinquished her dowry so that he could continue his education. Though penniless, she had married and now had adolescent children. It was Gregor who sent them through school.

Now, too, the grounds of the monastery were no longer restricted. Mendel previously had been limited to his tiny plot of land beneath the library window, but now he was free to plant his cherished peas—or the hawkweed that was his new interest—wherever he wished.

Mendel continued to breed and crossbreed the weed, but the work taxed his eyesight, which had always been poor. In 1869, he was forced to abandon it for a while. But now he was bewildered by the results that he had obtained; they were just the opposite of those he had arrived at earlier, when he had so closely observed the reproduction of peas.

Within a few years, he had completely dropped all study of the hawkweed. This was not the result of a conscious decision. Mendel simply paid less and less attention to the project —devoted less and less time to it—until it was almost forgotten. The spare moments he had he devoted to other forms of science, such as the meteorological studies with which he was fascinated, the observation of sunspots. But there were few such spare moments, for Father Gregor was saddled with administrative duties and then became embroiled in a long and bitter tax dispute with the government.

Gregor Mendel

Except for that tax dispute, Mendel's career was without a blemish. He had achieved a respectable place in his small community of Brünn, as well as a respectable income. It enabled him to travel occasionally, to go as far afield as Rome. He was appointed to a number of positions of importance, at least locally. When he contributed a large sum of money toward the formation of a fire brigade in his native town of Heinzendorf, he was made an honorary member. He was elected to the Agricultural Society. He became head of the Moravian Institute for Deaf Mutes and chairman of the council of the Moravian Mortgage Bank.

He was loved and respected, and at his death, on January 6, 1884, Mendel was mourned by the entire community of Brünn. He had been a kind and honorable man; he had lived a good life. But it was not until seventeen years later that his genius and his accomplishments were recognized.

The questions of heredity raised by Darwin's theory of evolution had not yet been answered. But throughout the world, scientists were engaged in research that would shed light on them.

As early as 1881, a German botanist, W. O. Focke, published a treatise in that field in which he mentioned Mendel's name with the highest praise. Focke had himself come across a reference to the priest in a scientific paper during the 1870s, although later he could not remember exactly what that work was. However, he found a copy of Mendel's paper and read it just before the publication of his own.

Nearly twenty years later, research on the problem had become more important and widespread. Throughout Europe, botanists were devoting themselves to such study, among them a professor at the university in Amsterdam, Hugo de Vries. In addition to conducting his own experiments, he searched out every available book, paper, and pamphlet that might be pertinent.

Reading Focke's paper, de Vries noticed the reference to Mendel and learned from Focke's bibliography that there was a copy of Mendel's paper in the library of the university at Uppsala, Sweden—one of those printed and distributed by the Brünn Society for Natural History. De Vries obtained it and recognized the validity and value of Father Gregor's experiments. When de Vries published two papers in March 1900—one in French, one in German—he gave Mendel due credit.

Just one month later, on April 24, 1900, a German professor at Tübingen, Karl Erich Correns, reported on his own work on heredity. Like de Vries, he had come across Mendel's paper. Now he, too, verified Mendel's results.

And in June of that same year, still a third scientist, the Austrian Erich von Tschermak, who had worked independently of the others, also presented a paper. Once again, tribute was paid to Mendel and the accuracy of his discoveries confirmed.

In homage to the man who had pioneered the work on which the modern science of genetics is based, the three botanists agreed that the three laws Mendel had postulated should thereafter be known as Mendel's laws. It was a fitting, but belated, mark of respect to the simple peasant's son.

Eventually, even the people of Brünn agreed to honor the ignored genius who had worked among them, by accepting the statue proposed as a memorial to him by scientific communities throughout the world and permitting it to be placed in the city's main square. There were few in Brünn who remembered the portly abbot, for it was then 1910 and forty-five years had passed since his brief moment of glory when he

Gregor Mendel

had read his paper before his colleagues in the Brünn Society, and twenty-five years since his death. In fact, some local citizens objected to the commemorative bronze. Would it not interfere with the erection of the booths for the local fair? But the townsfolk conceded at last, and there the statue of Father Gregor Mendel now stands.

Little was known of the life of Mendel, and even that little would have been forgotten had it not been for the ardor of another teacher of natural science in Brünn, Hugo Iltis. He had read Mendel's work in the public library of Brünn while he was still a student and had been impressed by it, although he had no more understood it than had a professor to whom he had shown it. But when the interest of the scientific world was focused on Mendel a few years later, the young man determined to search out what facts he could about the priest, consulting the records that remained, interviewing those still alive who might have known Mendel. In 1908, he began to publish articles on Father Gregor; in 1924, he published the only definitive biography of Mendel.

Iltis's publications have been of interest to the world, scientist and layman alike. But it is the modest village priest's discoveries that will perpetuate his name.

His research already has had a profound influence from a practical point of view. The determination of the negative Rh factor in the blood of newborn babies, and therefore the possibility of taking preventive measures to save their lives, is only one of many examples from the field of medicine. The great advances in the field of agriculture, with crops bred and crossbred to produce greater yields from each plant, as well as plants more resistant to disease, are due to Mendel's findings.

It is the theoretical aspect of Mendel's work, though, that may truly hold the key to the future. As the English botanist,

Scientist Versus Society

William Bateson, wrote in his own book, *Mendel's Principles of Heredity*, in which he included translations of Father Gregor's two brief reports, "Each of us who now looks at his own patch of work sees Mendel's clue running through it: whither that clue will lead, we dare not surmise."

V
SIGMUND FREUD

SIGMUND FREUD, LIKE Gregor Mendel, was born in Moravia, part of Austrian Silesia, although his name would forever be associated with his adopted city of Vienna. It was there that he was taken as a very small child, and there that he would live most of his life.

The Freud family, too, was beset by financial difficulties and poor by the standards of most of their neighbors. Yet, by comparison with the abject poverty of the peasant forebears of Mendel, they would have seemed extremely well-off.

Jakob Freud, Sigmund's father, was a merchant, selling wool to local textile weavers. He had married when he was only seventeen, and his young bride had borne him two children, both sons.

Her life had been brief, and Jakob Freud was a widower early. When he was forty, he married a second time. Jakob's older son, Emmanuel, had himself married by this time and had presented his father with his first grandchild, a son named John. When Sigmund was born on May 6, 1856, in the little town of Freiberg, he was already an uncle, a fact that was

Scientist Versus Society

to puzzle him throughout his childhood and to leave memories on which he would draw in later life in evolving some of his most important theories.

Although the family was a loving one, providing the tenderness and security in which a child could thrive, there were ominous warnings from without. For almost twenty years, Freiberg had been suffering from an economic crisis; unemployment among the weavers constantly increased and trade was almost ruined. There also was a developing spirit of nationalism among the Czech people, a more and more urgent need to break away from the culture to which they had long been subservient and to assert their own.

Frustrated in their desires, and ever more anxious about their monetary situation, the Czechs turned against the people who had been scapegoats frequently throughout history, the Jews. Being Jewish, Jakob Freud found it prudent, if not imperative, to leave the town for a less hostile environment. Vienna, the capital of the Austro-Hungarian Empire, seemed to fulfill his requirements.

An additional factor in Freud's choice was the large number of excellent schools in Vienna. There his children could be educated as he thought fitting.

Sigmund's mother, Amalie, was reluctant to settle in the distant city, but the prospects of the advantages there would be for her beloved first-born served to mollify her. She was certain that the boy whom she adored would have a remarkable future. He had been born with a caul, a thin membrane covering the head, which is usually broken in the process of birth. According to superstition, this was a sure sign of great good fortune. And an old peasant woman, seeing the child, had stopped her on the street to assure her that he was destined for fame and glory. Amalie Freud never wavered for a moment in her belief in her son's destiny, and her unswerving faith gave him the self-confidence that, in part, made possible the realization of her dreams.

The family left Freiberg in 1859, when Sigmund was only three, and spent a year in Leipzig before finally settling in Vienna. The Freuds had a second child, also a son, when Sigmund was nineteen months old, but he lived only eight months. In Vienna, the family grew rapidly, with five other children following in rapid succession.

Vienna was not at all what Jakob Freud had imagined. He did not prosper and once again was dogged by poverty. The living quarters he could provide for his brood were cramped and miserable. Only Sigmund was able to have a room of his own, where he could study undisturbed. His mother even gave him an oil lamp, though the other rooms were lit by candles.

And Vienna was no more free of anti-Semitism than Freiberg, although the forms it took were certainly less virulent. Nevertheless, a deep-seated prejudice against the Jews underlay the social structure, and they were tacitly excluded from a number of activities and positions. At times, the hostility broke into the open. At the age of twelve, Sigmund heard his own father's account of having his new fur cap knocked into the gutter by a young ruffian who ordered, "Jew, get off the pavement." When Sigmund asked, "And what did you do?" Jakob, to his son's shame, replied, "I stepped into the gutter and picked it up."

Although Sigmund himself faced few such overt acts, he suffered from the prevailing attitude throughout his school days and especially at the University where, as he wrote, "I was expected to feel myself inferior and an alien because I was a Jew." Later he found doors closed to him that swung wide open to men of far less ability.

This factor led to Freud's choice of medicine as a career. He had been a brilliant student at the *gymnasium*—the

Scientist Versus Society

equivalent of a high school—which he had entered at an age more advanced than that of most students, having been taught in his earlier years by his father. Like the others, though, he often dwelt on his future—that brilliant one predicted for him. Perhaps he would become a famous general, or even a minister of state.

But when it was time to enter the university, it was also time to face reality. And he knew only too well that the careers open to him, as a Jew, were limited. He could go into business or industry, or he could choose between medicine and law. There was nothing else.

Jakob Freud would perhaps have preferred that his son devote his energies to business, with its promise of financial security. But such a course was unthinkable to one of Sigmund Freud's intellectual bent. He seriously considered law but discarded the idea. Yet he at no time felt attracted to the physician's calling. He was, however, as he later wrote, "moved by a sort of curiosity, which was, however, directed more toward human concerns than toward natural objects." And it was after hearing a lecture on nature by the great German poet Goethe read aloud—an essay that emphasized not only the beauty of nature but its meaning and purpose—that he at last made his decision.

As a student, Freud's training was broad, including courses in philosophy as well as biology and zoology. In medicine proper, the only field that truly interested him was psychiatry.

By the end of his third year, he had embarked on several projects of original research. Those most satisfying to him, and which he performed superbly, were related to the nervous systems of animals of the lower orders.

In 1881, Freud earned his degree in medicine. But he still had no inclination to go into practice and was more than content to continue with the theoretical work that so engrossed him. This was carried out in the physiological laboratory of one of his professors, Ernst Brücke.

Although Freud was twenty-five years old, he still lived at home and was supported entirely by his father, except for an occasional small prize or token payment for an article published in a scientific journal. The following year, Brücke himself felt impelled to speak frankly to Freud on the subject of his future. There was no doubt about his ability, and no doubt, either, that eventually he could be appointed an assistant in the laboratory, thereby earning a modest salary. But such a position would not be vacant for many years. Had Freud been independently wealthy, he could have waited indefinitely for such an appointment. But with no financial resources whatsoever, it was essential that he abandon the career that so fascinated him and turn to the actual practice of the profession for which he had been trained.

Freud had an important additional reason for earning his own living now. He had recently met a charming young woman, Martha Bernays, and had fallen in love with her. He wanted desperately to propose marriage but was in no position to do so. Nor would he be, until he was sufficiently established to be sure of at least a modest income.

Freud therefore left the physiological laboratory and accepted a position as clinical assistant at the General Hospital, the leading hospital in Vienna. A short time later he was promoted, becoming a junior resident physician.

He carried out his duties in various departments of the hospital, working at first in the division of internal medicine. At last he was earning a small salary; he left home to live at the hospital. But he had no more interest in treating the sick than he had ever had.

After almost six months, Freud was able to transfer to the psychiatric clinic, which he found far more to his liking. There

Scientist Versus Society

he worked under a professor whom he considered a man of genius, Theodor Meynert. The focus was on the anatomy of the brain, and the assumption was that many otherwise inexplicable physical symptoms, such as sensory impairment, certain forms of paralysis, deafness, and blindness—all characterized by the term *hysteria*—were due to brain damage of some sort.

Freud became an expert at diagnosing organic diseases of the nervous system and at locating the exact spot of the injury so accurately that other physicians came from long distances, many of them from America, to attend his lectures. But those lectures ended abruptly after he ascribed the persistent headaches of a neurotic patient to chronic meningitis!

The young doctor was by no means alone in making such errors. The idea that such disturbances could originate in the mind—although the very term *psychiatry* is derived from Latin and Greek words that mean "the healing of the soul"—rather than in the brain, was almost unheard of. Freud had done his best to study nervous diseases, but there had been little chance to do so in the hospital itself, and much of the knowledge he gleaned on the subject he was forced to ferret out for himself.

Therefore, he had been extremely impressed by the approach of one of his older colleagues, Josef Breuer, who was to become one of his closest friends. Breuer had treated a young woman who was obviously suffering from hysteria, which had begun shortly after the death of her father. The woman—her name still appears in medical texts as Anna O.—seemed to go into a state of trance at frequent intervals. During those times, when Breuer was present, she spoke freely to him of the events that had preceded the onset of her various and numerous symptoms. The mere fact of talking about her troubles, of bringing them out into the open, seemed to relieve her terrible anxieties. Moreover, some of the signs of hysteria actually disappeared after these sessions. And Breuer was able to bring further solace to his patient by hypnotizing her and encouraging her to talk about whatever troubled her. This method, which came to be called catharsis, is still used in the treatment of mentally disturbed patients.

Breuer was not alone in exploring the use of hypnosis in the treatment of hysterics. In Paris, Jean Martin Charcot, at the Hôpital Salpêtrière, was conducting experiments on the same subject. Since the Middle Ages, hysteria, in most countries, had been assumed to be evidence of possession by the devil. But Charcot was able to induce all the classical signs of possession when he hypnotized his patients. Word of this had already reached Vienna and Sigmund Freud was profoundly affected by it. Before long, he had made up his mind that he would somehow get to Paris, to study with Charcot.

It was obvious that he could never reach that goal without help from others. He longed to marry Martha, yet knew it would be years before that was possible. Meanwhile, he lived on his miniscule salary, supplemented by what he could borrow. Josef Breuer, who had a lucrative practice, as well as a reasonably large fortune of his own, helped support his young colleague for many years.

Nevertheless, to reach Paris and fulfill his cherished dream of studying with Charcot, some sort of grant was necessary. And that grant would only be available to a doctor who had at least achieved the position of a lecturer in the hospital. With his usual single-mindedness, Freud devoted himself to attaining just that post. He was rewarded in 1885 with a traveling fellowship "of considerable value" and left Vienna for Paris.

The young doctor had already published a number of reports in his chosen field of neuropathology. But his interest also had been aroused, during the past three years, by the discovery of cocaine, a new drug that seemed to him miraculous. It could be used to relieve anxiety symptoms, and he frequently took it himself, when depressed, to change his mood. He urged it on others as well, recommending it especially to a friend who had become addicted to morphine, which he had begun to use to ease the pain resulting from a terrible operation.

Freud also recognized the anesthetic properties of cocaine, especially in the treatment and minor surgery of the eye, and suggested to another friend that he perform a number of tests with it. He himself dropped the matter in order to visit his fiancee, Martha Bernays, who was then living in Hamburg, and whom he had not seen for nearly two years. Others took up the work, demonstrating its efficacity—and gaining credit for the discovery, which would have made Freud famous at an early age had he continued his own work.

While the use of cocaine as an anesthetic was of tremendous value, the dangers of the drug for other purposes soon became apparent. Freud's friend, who had been cured of his addiction to morphine by the substitution of the new narcotic, now found himself a slave to it. Throughout Europe, others had also fallen victim to the habit-forming properties of cocaine. The problem had become so widespread that one physician described it as "the third scourge of humanity." Freud's advocacy of cocaine was to be used against him later, in an attempt to discredit his findings in the field of psychoanalysis.

But cocaine and its pernicious effects were far from Freud's mind during his stay in Paris. He was deeply involved in his own studies there but disappointed that he had no personal contact with the great man whose fame had attracted him to the city. This was remedied when Freud wrote to Charcot, asking permission to translate a series of his lectures into

German. The offer was accepted, and from then on Freud enjoyed special privileges in Charcot's clinic, as well as the friendship of the older man. He was able to describe to him Breuer's work with Anna O., which seemed so closely connected with Charcot's own theories. The Frenchman, however, showed no interest in the case, a lack of comprehension that bewildered Freud.

Freud left Paris in the spring of 1886. On his way home he spent several weeks in Berlin, studying the diseases of children in order to prepare for the post offered him as the director of a new neurological department in the first public institute for children's diseases in Vienna. It had been proposed to him before he left for Paris and Freud had accepted, knowing that, as a Jew, there would never be a place for him at the university's psychiatric-neurology clinic.

In addition to the thrice-weekly sessions at the institute, the young doctor accepted private patients. These were few and far between, and his prospects of marrying his beloved Martha seemed to diminish daily. Freud was practically penniless. He earned a pittance and felt obliged to share even that with his parents. Martha had a very small dowry, but it would be needed for the meager furnishings of the new home—for such mundane items as linens and pots and pans, along with such essentials as a bed, a table, and a few chairs.

At last, however, the couple had a stroke of marvelous luck. Martha's wealthy aunt in Brünn sent the pair the munificent sum of $500 as a wedding present. There was half as much again from an uncle in London. After an engagement that had lasted five interminable years, the two could finally wed.

Even so, there were complications. Martha came from an orthodox Jewish family; Freud, although Jewish by birth, was an atheist. The idea of being married in a religious ceremony was almost intolerable to him, but under Austrian law a civil service alone did not suffice; a religious ceremony was required as well. Sigmund and Martha, therefore, submitted to both. The two ceremonies were performed on September 13, 1886, and after a brief honeymoon the couple settled down to life in Vienna.

A little more than a year later, their first child, a daughter, was born. Five other children came quickly after her. Freud was devoted to them all, a kind and loving father, as he was a kind and loving husband. But he was haunted by the specter of poverty. His earnings were small, the family was large, and there simply was not enough money to go around. When Martha gave Sigmund a necktie, making him the proud possessor of two, it seemed that he was rich beyond his wildest dreams.

Freud supplemented his small income from his work at the institute and from his few private patients with his occasional writing. He published several papers on neurological disturbances of children, based on his treatment and observation of them. The articles were almost completely disregarded at that time, although they are now considered classics in the field.

He published a book on the subject, too, which also was ignored. Of 850 copies printed, 257 were sold, and after nine years the rest were destroyed.

With private patients, Freud at first employed the usual methods of his time. The results were far from gratifying, and at the end of 1887 he began to use hypnosis. By this means he found that patients were often able to dredge up memories so painful that they had repressed them—the word was Freud's own and has since entered the analytic vocabulary.

The new hypnotic method was far more effective. Freud at last had the impression that he was truly helping his patients. Nevertheless, there were times when he was unable to induce a trance sufficiently deep to release the suppressed and disturbing ideas. Moreover, the results were rarely lasting.

Searching for a new approach that would retain the benefits of hypnosis while replacing the elements that seemed useless, Freud encouraged his patients to speak to him, much as Breuer had with his cathartic method. He discovered that if the patient lay down with the psychiatrist's hand on his forehead at first, some sort of self-hypnosis was induced and again the painful memories surfaced.

In the beginning, Freud questioned his patients, sometimes making suggestions to them as well. When one woman complained that he was interrupting her train of thought, he stopped the practice, and allowed her and the others to say whatever came into their minds. This process—now known as free association—is the basis of modern psychoanalysis.

A great number of Freud's patients were hysterics, victims of the same strange malady that had first drawn the doctor to Charcot. Most physicians considered it a malady of women only—the word hysteria itself comes from the Greek term for uterus and from ancient times its cause was considered to be a disturbance of that organ. But Freud, working with Breuer, discovered cases of hysteria in men as well.

Further, he had an important insight into the nature or at least the cause of the illness, one that Breuer never was able to accept. In case after case, there was definite evidence of sexual maladjustment of some type. In 1895, the two men published a book together, *Studies on Hysteria*. Freud made important contributions to it, but insisted that the main ideas were those of Breuer. Certainly, the sexual basis of hysteria was omitted.

Three years earlier, Freud had begun to question his patients without the use of hypnosis, moving gradually to the now accepted technique of free association. At first, the torrent of words that spilled forth seemed unintelligible, incoherent. But gradually they began to form a pattern. To Freud, the memories of dreams seemed especially fruitful in shaping that pattern.

Two years after the publication of *Studies on Hysteria*, Sigmund Freud set out alone to perform a remarkable feat. It was nothing less than his own psychoanalysis.

He was prompted in part by his desire to gain insight into the workings of the human mind, in order better to help his patients. But the main impetus came from the death of his father, to whom he had been deeply attached. He wrote that all his early feelings for his father had been reawakened and "I feel uprooted."

The magnitude of the task was formidable. In the classical analytical situation, there is a dialogue between the analyst and the analysand (or patient). The former—through his or her training analysis—has explored the way before, has experienced the same emotions the patient experiences. Moreover, the very presence of the analyst eases the terror the afflicted person is sure to feel, just as a companion on a dark and lonely road brings comfort and reassurance, quieting, at least to a certain extent, otherwise unbearable fears.

Even under the best of circumstances and with the most competent and sympathetic help, many neurotics are unable to complete analysis, or even to begin it. The memories that they repressed earlier, because of the terrible pain associated with them, are still too agonizing to contemplate. Many others embark on analysis in an effort to obtain relief from their present sufferings, unaware of the anguish in store for them.

Freud knew only too well what lay ahead. Yet he probed the depths of his unconscious, facing his fears alone, getting to know himself as he really was, with all the flaws and ugliness that characterize every human being. Although the major part of his self-analysis was completed within the next few years, he continued this form of exploration for the rest of his life.

Sigmund Freud

During the period in which Freud was searching out the secrets of his own soul, he also was engaged in writing the book that he always would consider his most original work, *The Interpretation of Dreams*. It was published in 1900, in an edition of 600 copies; it was more than eight years before all were sold.

In his own analysis, and in his work with others, the subject of sexuality continued to appear in some form. Especially in the treatment of hysterics, Freud found what seemed to be a common theme, a common memory. Almost without exception, as hysterics remembered the past, they divulged an attempt on the part of the parent of the opposite sex—usually it was a woman speaking of her father—to commit a sexual assault of some type on them.

For many years, Freud believed implicitly in what he was told, although the number of apparently perverted fathers in Vienna taxed even his credulity. Eventually, though, it dawned on him that the idea of seduction was a fantasy—a wish—on the patient's part. It was not the child who was the innocent victim; it was the parent who was the object of an incestuous desire.

Freud's own analysis confirmed this idea and led to the development of his Oedipal theory. He took the name from the Greek legend of Oedipus, the son of Laius and Jocasta, the king and queen of Thebes. In the myth, their child was abandoned as an infant. He returned eventually and, ignorant of his identity, killed his father and married his mother. The Oedipus complex thus took into account not only the erotic attraction of the parent of the opposite sex for a young child but his hostility toward the parent of the same sex. In the normal person, these attitudes disappear at an early age, but in the neurotic they persist.

The importance of the repressed sexuality in the hysteric led the still puzzled scientist to explore the same notion with those whom he described as "the so-called neurasthenics" who crowded into his office. As he did so, however, the crowds thinned out until there were only a few individuals left. The years in which Freud was making his most important discoveries were the height of the Victorian era. The very subject of sex was taboo; even proper anatomical terms were beyond the pale. Legs had become limbs. The word breast was so shocking that chickens developed white meat and dark.

Yet Freud persisted. He had come across a truth, and as a man of science he was impelled to pursue it. The basis of neurosis was a sexual malfunction. Moreover, dreams expressed the repressed desires, the emotions that were unacceptable in a waking state.

It was not only the layperson who was shocked by the new discoveries. Even Freud's old friend and collaborator, Josef Breuer, broke with him because of his ideas. At a meeting of the College of Physicians in Vienna, when the analyst attempted to explain his views of the sexual origins of neurotic symptoms, he was both pleased and relieved to find that Breuer supported him. As the two left the meeting later, Freud thanked the older man, telling him how grateful he was for his support. Breuer stared at his former student with mixed despair and disgust. At last he turned away, snapping, "I don't believe a word of it."

If Breuer believed not a word of what he had heard at that particular meeting, he believed even less of what he read in Freud's next important work, *Three Essays on the Theory of Sexuality*. Freud scrutinized the sexual needs and desires of infants, tracing their development from the earliest years to adulthood. Moreover, Freud recognized and expressly stated the dominant instinct of aggression, even in the very young.

Once again, Freud's ideas were far afield of those of his contemporaries. The very thought of infant sexuality was too frightful to contemplate; hostility and aggression were almost as bad. If they were rejected out of hand, as they were, it was not merely for their novelty and their inherent content; it was also in the context of the times.

The Victorian era had spawned a widespread prudishness and puritanism. Moreover, it had evoked an aura of romanticism that seemed to envelop all.

Much of the world lived in appalling squalor and misery. Cruelty was commonplace. But the well-bred, the educated members of the upper classes whose viewpoint shaped the prevailing mood, preferred to look at the world through rose-colored glasses. The English poet, Robert Browning, summed up their attitude in the lines:

> God's in his Heav'n
> All's right with the world.

Darwin, with his now widely accepted theory of evolution, had not only shown the descent of humans from the most primitive of the species; he had also shown humans *ascending*. It was not only the *fittest* who survived but, in a moral sense, the *best*.

War, except for colonial conquests, was a thing of the past. The invention of dynamite by the Swedish industrialist Alfred Nobel had simply made it too terrifying. Or so it seemed, just fourteen years before millions were slaughtered in a global conflict, just forty years before the Second World War.

In that best of all possible worlds—the phrase Voltaire had used ironically two centuries ealier—children were perfect. Childhood was a period of the most complete innocence, of purity, of happiness. And suddenly a Viennese Jew had appeared to dispute that irrefutable truth.

Until that time, Freud had been ignored for the most part. Occasionally, in England, in America, even in Australia, a short review of one of his books or papers appeared. Occasionally, it was even sympathetic. And from time to time there was a letter from such far-off places asking for further details, for explanations, for information about treatment for one of the many disturbed persons begging for help.

Then in 1905 Freud published an actual case history, that of a young woman whom he called Dora. And now the storm that had been brewing over his head burst in full fury. The least offensive criticisms were those that appeared in print; when writing even about matters to which they are deeply opposed, scientists are likely to consider seriously the laws concerning libel. But they were not so restrained at scientific meetings, where their remarks were off the record.

Freud and his few followers, they insisted, were all sexual perverts. Freud's books were far from being scientific works, but mere pornography. His methods were even worse, involving the most obscene approaches. When the subject of psychoanalysis came up at a meeting of neurologists and psychiatrists in Hamburg, an eminent professor banged his fist on the table, shouting, "This is not a topic for a scientific meeting. It is a matter for the police."

In Vienna, Freud was snubbed on the street and often verbally abused. And yet he had managed to attract a few disciples, some of whom remained faithful to him and his ideas to the end, while some were later to break with him.

Freud had long wanted, had needed, some professional post, some official recognition. He would have liked to be named professor at the University of Vienna. He learned that he had actually been proposed as an associate professor there, but approval was out of the question. There was no place there for a Jew—and most certainly none for one with his views on sexuality.

Sigmund Freud

Finally he came to the conclusion that he was enduring martyrdom without reason and unnecessarily. There were ways and there were means, and while they might be slightly less than honorable, they were still universally accepted. And so, through some of his patients, he made the contacts that were imperative to his appointment. One eventually traded a painting accessible to her for the position Freud longed for—that of "Professor Extraordinarius." Nothing was paid to him and nothing was required of him, but he was at least allowed to give lectures at the university, which he did twice a week for more than ten years. Moreover, it was now permissible to call him *Herr Professor*, a mark of respect in Vienna.

Another mark of respect was the attendance of several Viennese doctors at Freud's lectures on the psychology of the neuroses. Eventually the few who were most interested decided to meet once a week at Freud's apartment to discuss his work. The small group expanded as foreigners came to consult the man who had achieved such remarkable success in dealing with formerly incurable mental illnesses.

These foreigners were to bear the brunt of the hatred and invective spewed forth against the cause of pyschoanalysis. The first victim was an Australian clergyman, Donald Fraser, who was forced from his post as a Presbyterian minister because of his support for the unorthodox ideas of Sigmund Freud. A short while later, Ernest Jones, who was eventually to write the definitive biography of his friend and teacher, lost his place in a London hospital for encouraging his patients to discuss their sexual acts and fantasies with him.

In Berlin, the same fate befell another doctor who also employed the new method of treatment. A Swiss director of a seminary was dismissed when it was learned that he accepted

Scientist Versus Society

Freud's view, while in Sweden the career of a noted scholar in the field of linguistics was ruined when he was denied the right to lecture at the university after he had published an essay elaborating on a Freudian theory on the sexual origins of speech.

Despite the persecution of the early analysts, the movement continued to attract adherents. They came from all parts of the civilized world to study with Sigmund Freud and to undergo a brief training analysis with him. Then they returned to their own countries, to treat the mentally disturbed seeking relief from crushing emotional problems.

Small but important analytical associations sprang up in many places. The most prominent group was that in Switzerland, which drew almost all its membership from Zurich. It was led by C. G. Jung, who served Freud's cause well at first. Later he broke with Freud, denying Freud's fundamental ideas and substituting for them his own theory based on mythology.

In Budapest, Sandor Ferenczi was Freud's most distinguished supporter. He too, would eventually break away from Freud as his own views developed and changed.

Princess Marie Bonaparte translated many of Freud's works into French and led a society in Paris. Unlike many others, she remained loyal to him throughout her life. She helped him financially, giving generously from her personal fortune. And many years later she was able to use her position as a Greek and Danish princess to spirit away for safekeeping —under the very noses of the Nazis—his collection of manuscripts and papers.

The greatest enthusiasm for psychoanalysis—Freud had first used the term in 1896—was that aroused in America. In 1909, both Freud and Jung were asked to give a series of lectures at Clark University in Worcester, Massachusetts. At that time, an honorary degree was awarded Freud—the only one he ever received.

Patients had begun to come from America, and now Freud spent much of his time treating them. His greatest period of creative work was past, yet he continued to write and publish. Many of his papers appeared in the psychoanalytical revues that mushroomed as the movement expanded. And many of these works were remarkable for their depth of thought.

Among them was *Totem and Taboo*, first published in 1914. In this work, Freud applied the principles of psychoanalysis to anthropology. He published another book in 1920, the most controversial of all his works, called *Beyond the Pleasure Principle*.

This work made two major contributions to analytical theory. One was the concept of the compulsion to repeat. The second was that of the death instinct.

Much earlier, Freud had defined a principle of pleasure and of pain—or "unpleasure," as he called it. In its simplest form, unpleasure was merely the stress that resulted from overexcitement, while pleasure was its resolution or ending. The whole theory of wish fulfillment, as expressed in dreams, was based on just such a proposition. A repressed desire brought about tension and anxiety; when it was satisfied in a dream, the anxiety disappeared.

But now he noticed, and noted, examples of the acts or dreams that were constantly repeated. They were always painful, and instead of wiping out strain, they increased it. These repetitions, he decided, came from a deep and instinctive drive. The purpose of this drive was to return to the state that existed before the one when pain and anguish first appeared. That state, ultimately, was death.

However, Freud at no time intended his concept of death instinct to be interpreted—as it has been so often—as a

Scientist Versus Society

suicidal impulse, a death wish. Death arrived at a definite and predetermined time, and it was the nature of humans—with their instinct for self-preservation—to survive until that appointed hour.

It was, in fact, their instinct to survive beyond that moment. Physically, of course, this was impossible—the soma or body inevitably would die. But the psyche or soul could continue its existence in one generation after another. Neatly, Freud bound in his theories on sexual drives with those on death.

Beyond the Pleasure Principle was the only one of Freud's many books to take on a metaphysical tone, at times leaving the realm of science to soar into philosophy. It turned out to be his least accepted work as well. Few of his followers could agree with his premise, and even today the psychoanalytical movement is split between the minority for whom the death instinct is an essential part of the human spirit and the majority which has discarded the notion.

His opponents within the analytical movement attributed the new theories to Freud's many personal tribulations. The terrible First World War was over, and the two sons who had gone to the front had now returned. But the anxieties of those days had been almost unbearable.

Peace had brought with it new problems, and Austria, like Germany, had sunk into a disastrous economic situation. Inflation was rampant, with the cost of even the most basic necessities soaring daily. The food available was barely sufficient to sustain life; often there was no fuel to heat Freud's consulting room and he received his patients wearing a heavy but threadbare overcoat and thick woolen gloves. After the last patient had left, he turned to correcting proofs of books and papers with frozen fingers.

His savings had been wiped out and there were no longer patients from Vienna or the rest of Eastern Europe, as there

Sigmund Freud

had been before. There were, however, those from England and America—sent by the faithful Ernest Jones—who could pay in still valuable dollars and pounds. There were gifts, too, from Martha Freud's brother, now a successful businessman in the United States, and loans from a few followers. Although in 1919 Freud was at last made a full professor at the University of Vienna, the title was an empty one, giving him neither teaching duties nor a seat on the Board of Faculty.

The following year brought tragedies—first, the loss of his close friend Anton von Freund, and then his beloved second daughter, Sophie. She was only twenty-six when she died of influenza.

For a while, it was only his indomitable courage—which had sustained him through his torturous self-analysis—that kept Freud going. Gradually, though, the tide seemed to turn, at least as far as his work was concerned. The psychoanalytical societies that had struggled for mere existence in the years before the war had suddenly expanded in a spurt of activity. Moreover, psychoanalysis had become almost respectable!

The change in attitude was a direct result of the terrible war. People everywhere had behaved like savages, and among the victims of the carnage were all the romantic notions about the inherent good of the human animal. The time was ripe for a new look at human beings and their motives, and there seemed no better starting place than with the theories set forth by Freud during the preceding quarter of a century.

Gradually, the hardships to which Freud had been subjected eased. He began writing again, completing two new books, *Group Psychology* and *The Ego and the Id*. The latter appeared in 1923, when its author was sixty-seven years old.

He had suffered from poor health throughout much of his life. Often his illnesses were psychosomatic—induced by emotional strain. Now, though, it was discovered that Freud had cancer of the jaw.

An operation was performed to remove the malignancy—one so simple it was supposed that hospitalization would not be necessary. But the surgeon, either through lack of skill or sheer negligence, bungled it so badly that Freud nearly bled to death. Furthermore, the normal precautions to prevent the scar from shrinking were not taken, so that eventually the opening of the mouth was greatly reduced.

That was the first of thirty-three operations, and Freud suffered almost constant pain from that day on. He was forced to wear a huge denture called a prosthesis in order to speak or eat. It was so difficult to insert and so uncomfortable that in the family it was known as "the monster."

In 1923 came another event, little noticed at the time, that was to have a profound effect on the life of Sigmund Freud. It occurred in Munich, where an unemployed housepainter named Adolf Hitler gathered a group of thugs in the basement of a beer hall and poured out all his hatred, all his rage against the Jews. They were, he said, to blame for whatever had gone wrong. If there were troubles in the postwar era—and they were already too numerous to count—it was because of that despised race. Hitler pledged not only vengeance on the Jews but the restoration of the defeated Germany to what he considered its rightful place in the world.

When he attempted to carry out his plans by taking over the Bavarian government, he was quickly arrested, tried, and imprisoned. He spent his time in jail writing *Mein Kampf*. Now considering him a martyr, the discontented Germans flocked to join his National Socialist—Nazi—movement.

The ominous rumblings from Germany swelled to a roar during the following years, but Freud was preoccupied with his own problems. As much as possible, his physical ones were soothed by the devoted attendance of his youngest daughter, Anna, who served as his sole nurse. She was later to become a noted analyst, working mainly with children.

In spite of the pain, Freud continued his work, both in treatment and in writing. He had to, in order to pay the medical bills that piled up on his desk. And once again there was dissension among his followers. Otto Rank, a Viennese doctor who was then lecturing in New York, had suddenly abandoned his teacher's theories to propose his own—that all neurosis was the result of the trauma or shock of birth.

Ferenczi was causing trouble, too, by advocating lay analysis—that is, analysis by a person who was not a physician. In doing so, he seemed to give official recognition to the numerous quacks and charlatans who had appeared. In addition, he had developed a new method of analysis. In this approach, he would abandon the distanced and objective stance of Freud, permitting such physical contacts as kissing and caresses, if the patient desired them.

Such a proposal naturally brought back to the front the early objections to psychoanalysis, when doctors in Hamburg had stated: "Freud's methods are dangerous because they simply breed sexual ideas in patients."

There was little for Freud to do but to write Ferenczi a remonstrating letter. The defection of two followers was a bitter blow. Yet it was softened somewhat by the belated homage paid him by the city of Vienna when he was given the keys to the city.

Three more books appeared, written under the most difficult circumstances. The most famous, published in 1930, was *Civilization and Its Discontents*, which described the endless conflicts between the desires of the individual for personal satisfaction and the prohibitions placed on him by society for the collective good.

Scientist Versus Society

Freud's health continued to deteriorate. He now had a heart condition that required treatment. In addition, an old intestinal complaint recurred. Both made it impossible for him to accept two significant tributes that had at last come his way. Freud was unable to deliver his own address when he was awarded the Goethe prize by the city of Frankfurt in 1930. In his place, his daughter Anna read it for him. And in 1931, when he was the first German since 1898 to be invited to give the annual Huxley lecture at the University of London, illness forced him to decline.

He was honored, too, by his native city of Freiburg, on his seventy-fifth birthday, when a bronze plaque was placed on the house in which he was born. Although he was pleased, he remarked sadly that he would gladly have traded the recognition for "a bearable prosthesis, one that didn't clamor to be the main object of one's existence."

The Goethe prize was the last mark of respect to be shown the now aged savant by Germany. That country, like Austria, had been plunged into the depths of a depression. The economic problems brought in their wake the most grave political ones. And now, to many Germans, the one hope of salvation seemed to lie with the Nazis.

Freud remained optimistic, however, contending that no country that had produced a Goethe could ever accept a regime that made a national virtue of racial hatred, of oppression, of the denial of all the human rights so painfully won throughout the centuries. Further, although many analysts had already fled Europe, he was convinced that the horrors becoming more and more frequent in Germany would never spread to other European countries.

He was wrong, of course. In 1933, Adolf Hitler seized power in a putsch, and what had seemed before a terrible nightmare became a reality. One of the fuehrer's first acts was the public

Sigmund Freud

burning of books that were offensive to the new leaders. These included works by Jewish authors as well as those expressing ideas—the Nazis termed them decadent—that conflicted with the concepts of the Third Reich.

Freud's works qualified on both counts, but he accepted the fact with equanimity. It was a sign of progress, he commented, that only his books were being burned; a thousand years earlier, he himself would have gone to the stake.

Once again, he was wrong. Had he lived and stayed in Vienna, he might have shared the fate of his four aged sisters, who were exterminated in the Nazi gas chambers.

Signs that the terror was spreading were everywhere. Yet Freud insisted on remaining in the city that had so long been his home. It was only after the Anschlüss, when Austria was annexed by Germany, that he was at last persuaded to leave for England.

Safe passage had been arranged for him and his family, along with two servants and his personal physician, through the intervention of William Bullitt, then the American ambassador to France and also an old friend. Bullitt managed the difficult feat by an appeal to President Franklin D. Roosevelt.

Before they could leave, Freud's son Martin had been arrested a number of times. His daughter Anna was also arrested by the Gestapo and questioned throughout a whole day, although she was not tortured. At the time of their departure, Sigmund Freud was forced to sign a paper stating that he had been treated by the German authorities, and particularly by the Gestapo, with the greatest respect, that he had been completely free to continue his scientific work without interference, and that he "had not the slightest reason for complaint."

Freud made no protest but insisted on adding one sentence

of his own. In an ironic note he wrote, "I can heartily recommend the Gestapo to anyone."

In spite of his age and his failing health, the journey to England was made without incident. There Freud was warmly welcomed. In 1936, on his eightieth birthday, he had been elected a corresponding member of the Royal Society, and now British intellectuals did their best to make his life in his newly adopted country as easy as possible. They had, in fact, already prevailed upon top government officials to sign the papers he needed in order to reside in their country and to continue his work if he, or Anna, chose to do so.

Freud had brought with him the manuscript of *Moses and Monotheism*, the last of his important works to be published during his lifetime. Settled down in a comfortable house in London, overlooking a garden that gave him much pleasure, he undertook the treatment of a few patients. But most of his time was devoted to writing *An Outline of Psycho-Analysis*, summing up his work.

He was never to finish it.

The fearful pain Freud had suffered for so long continued, always seeming to increase. At last it became unbearable. Then, and only then, he reminded his friend and physician, Dr. Max Schur, of a promise made to him: that "you would help me when I could no longer carry on."

The next morning, Schur gave Freud a third of a grain of morphia. He was unaccustomed to any drugs—only toward the end had he accepted even aspirin—and the morphia brought a deep sleep from which he never awakened.

Freud died on September 23, 1939, at the age of eighty-three. His body was cremated, and the ashes placed in an ancient Greek urn that had been a gift from Marie Bonaparte. A number of old friends attended the brief rite that followed at Golders Green Cemetery in London.

Freud was gone. But he had left behind a new, deeper understanding of the nature of man.

Even now, there is dissension about his work, as there was during his lifetime. But his influence has been enormous. Scarcely any field of human thought today does not take into account the truths Sigmund Freud discovered.

VI
CHARLES DREW

To be a peasant in Moravia meant grinding poverty and an almost total lack of opportunity to rise to a better station in life. Being a Jew in Vienna meant humiliation. Being black, in America in 1904, very often meant both.

Less than fifty years before, slavery had still existed in the United States. It would take another fifty years before the Supreme Court, in the historic decision known as *Brown v. Board of Education of Topeka*, would declare that the doctrine of separate but equal in regard to schooling was unconstitutional.

Separate, of course, meant segregated—and for the most part, the schools reserved for blacks were far from equal to those for white students. Certainly, there was no equality whatsoever when it came to work or the chance to enter a profession. Only menial and manual labor was open to the father of Charles Richard Drew, in Washington, D. C., where the child was born on June 3, 1904.

Charles was the first of a large brood. By dint of hard work and stringent economy, the hard-pressed parents always

Charles Drew

managed to feed the family. But they could provide little else. And the youngsters were obliged to do their share, too. Charles was no more than seven years old before he was out peddling newspapers on the streets of Washington, adding what he could to what his father and mother could provide.

But there was play for little Charley, too, or at least an active and constant participation in sports. There also were plenty of trophies to prove that he was a winner. All were proudly displayed in the small room he shared with his brother in the steaming tenement where the family lived.

Charles was only five when he placed first in the swimming tournament held in Washington on July 4—for Negro youngsters, of course. Black and white did not mix, most certainly not in swimming pools or at public beaches.

He'd won the tournament again the next year, and the next. And then, when he was eleven, he won the junior championship.

Swimming was only one aspect of sports in which Charles Drew excelled. When he was older, at Dunbar High School, he played football, basketball, and baseball, winning letters in all three. He'd been on the track team, too. With such a record, he was awarded a coveted prize, the James E. Walker Memorial Medal, as the school's outstanding all-around athlete. The fact that he'd been president of his class added to his laurels.

Dunbar was a school for blacks only. It had been named for the famous and talented poet Paul Laurence Dunbar, and was in itself a coveted prize since it was far superior scholastically to most of the schools in Washington that were open to Negroes.

Many boys from Dunbar went on to college. Most chose to enter Howard University, right there in Washington. Howard, a federal institution, had developed from a small theological

Scientist Versus Society

college into an important university. Since it had no color bar, the great majority of its students were blacks.

Others from Dunbar went to some of the private Negro schools or the land-grant colleges that were scattered throughout the South. They were good, but hardly first-rate, suffering perpetually from a lack of the funds needed to buy books, to equip laboratories properly, and especially to attract the fine and dedicated teachers who made many of the Northern white universities truly great places of learning.

With Charles Drew's brilliant record at Dunbar, however, it was possible not only to aspire but even to be admitted to one of those leading colleges. A scholarship would be necessary, of course. The kind of place Drew had in mind was far beyond his family's means.

But a scholarship was not out of the question. Only a few years before Drew was to apply, Paul Robeson, a superb Negro athlete and superb scholar, had been admitted to Rutgers University in New Jersey. His career there had created something of a sensation; besides being an all-American football player, he had been elected to Phi Beta Kappa, the national scholastic honorary society. He was later to have an immensely successful career as an actor and singer. Two other blacks had attended Rutgers. And if they had . . . well, what had Charles Drew to lose by applying either there or to an equally good white college, aside from the cost of postage stamps?

Robeson, however, had suffered cruelly from racial prejudice during his college days. At football practice, his teammates often kicked him viciously as he lay on the ground after he had been tackled. They dug their metal cleats into his bare hands. They threw their full weight against him although he was already down, the play already over. Robeson often limped off the football field when practice was over, bruised and bleeding.

Charles Drew

Drew had himself been humiliated often enough as a child when he had ventured into white neighborhoods. The epithet "nigger" had been flung at him; bigger boys had shaken their fists menacingly in his face.

He had no desire to face more of the same torment. And he knew that it was needless. The sons of the wealthier Washington Negroes were often sent to the private preparatory schools that had long existed in New England. There was a liberal tradition there, and a far more tolerant attitude than in other parts of the country. Colleges in that region had the same spirit.

And so Charles Drew sent out his letters of application to a number of small but distinguished Northern schools. He was accepted by Amherst, one of the best, which nestled in the rolling hills of Massachusetts. The proud possessor of a scholarship, he entered Amherst as a freshman in 1922.

Prior to that, Amherst had not been particularly noted for athletic excellence. But that year marked the beginning of a new era when the school forged to the front in collegiate sports.

It was there that Drew's great interest lay. Without a moment's hesitation, he went out for each and every team, making them all. He played baseball and football, and basketball as well. He was even elected captain of the varsity track team.

Winning is important to every athlete, the goal of every game. But it was especially important to Charles Drew, who seemed to find in it a compensation for the feeling of inferiority that haunted him, the awareness that he was different from his fellow students.

As he had expected, he was not shunned because of his race. But he was only too conscious of the great gulf between his own background and that of the others. He was, to begin with, poor—and poor in a school that catered to the sons of the wealthy. His clothes were often shabby, his shirt collars frayed, his trousers threadbare or even patched.

Drew felt uneasy, too, about his earlier schooling. True, Dunbar was an excellent high school with an enviable reputation. But it could never offer its students the special advantages of the renowned private schools from which such a large part of the Amherst student body was drawn—places like St. Paul's in Concord, New Hampshire, or Philips Exeter or Choate or Groton. Classes were small in such places, a great boon since it meant far more individual attention than was possible in a huge public high school. And the boys who attended those schools almost always came from highly cultured homes. They had listened to the best music all their lives, had extensive libraries in their own homes, had been introduced at an early age to great works of art. Many of them had traveled widely. Above all, they were young gentlemen, secure in the knowledge of their social position and therefore very much at ease in almost any situation.

Although Drew was well-mannered and unfailingly polite, he could hardly help comparing—and to his disadvantage—his own heritage with that of the others. And so he set out to do supremely well in what he did best—to excel in athletics.

He succeeded. In each of the four years he spent at Amherst, he won the Cobb Pentathlon Award for his achievements in five different sports. But, at least during his first year, it was at the expense of his studies. He was about to fail chemistry when he was called in by the dean for an explanation. There was none—at any rate, none that was valid. From that time on, Charles Drew devoted himself with as much fervor to his books as to athletics.

With a new perspective, he found that he had time for both.

His grades inched up to passing and then well beyond. They stayed there, too. Yet Drew won the Cobb Award again and again, as well as a mention as an all-American halfback. Then came what might have been a great triumph; it turned into one of his most bitter disappointments.

The world was preparing for the quadrennial Olympic Games and athletes throughout America, as in most countries, were competing for places on their national teams. The top track star from Amherst would be a likely candidate for the U.S. team and Drew hoped for the chance to show his mettle in the international arena. He had a rival, though, another student whose prowess was equal to his own. At last a final and decisive race was arranged between the two.

Both young men waited, nerves taut, for the starting gun, then were off with all the speed they could muster. It was Wilder, the second runner, who pulled ahead and stayed there for what seemed an eternity to Drew. Then, with a sudden surge of energy that came from sheer determination, Drew pulled up even. He stayed there, neither youth able to break ahead. And so the race ended in a dead heat.

Rather than asking the two exhausted athletes to run a second race, the coach decided to choose the winner by a flip of a coin. It turned out to be Wilder, by a stroke of fate. And so he went to the Olympic Games while Charles Richard Drew stayed at home.

He was back at Amherst the next fall, leading the football squad to one victory after another. In one crucial game, with Amherst the underdog, he seemed to outdo all his earlier efforts. The crowd cheered him on, which was at least some compensation for the pain that racked his body. That was part of the game, wasn't it?

But as Charley Drew went over the goal line for the final

and winning touchdown, the pain was unbearable. Just before he lost consciousness, he saw the blood streaming from a wound in his thigh where he had been accidentally spiked.

When Charles regained consciousness, he was in a hospital. He lay there for days, with nurses and doctors hovering over him. In spite of their care, an infection had set in, one that stubbornly refused to respond to treatment. As time passed, Drew's father was sent for, and Charles was dimly aware of the older man's presence at his bedside. Only later did he realize how serious his condition had been—that his very life had been in danger.

But he had the will to live, and his dedicated doctors had the skill to save him. And during those endless days Drew had the time to think of his own future. It wasn't long before he decided that he would go into medicine. The satisfaction of saving lives was sufficient reward for all the difficulties that lay ahead.

Drew was back at his familiar exploits in athletics before long. He graduated from Amherst with still another honor, the Mossman Trophy, given to the man whose achievements, both in sports and scholastically, had been most outstanding during his entire four years.

With a bachelor of arts degree in hand, Charles's next step was to earn a degree in medicine. Howard University seemed the logical place to continue his studies. Attending a local school, he could live with his parents and avoid the expenses of room and board that would be necessary if he went farther afield. Such a saving would be considerable; it was also imperative, considering the cost of medical training.

Just how high that was, however, Drew had never even imagined until the day he went to see the dean of admissions at Howard University in Washington. He discovered that Howard was far beyond his means, under any circumstances.

Charles Drew

There was a younger brother ready for college and Charles could hardly deny him his chance at an education in order to further his own.

Pushing the idea of the advanced degree to the back of his mind, he accepted a position as athletic coach at Morgan College, a school for blacks in nearby Baltimore. Two years later, when he had set aside the funds he would need, he confidently applied for admission to the Howard University School of Medicine . . . and was turned down!

His record at Amherst had been excellent, his grades far above average. Unfortunately, it was explained to him, he lacked several essential credits in English. Therefore, and regretfully, Howard could not accept him as a medical student.

It was a harsh blow, but one from which Drew quickly recovered. There were other universities with medical schools. And better ones than Howard. There was, for example, Harvard University in Cambridge, Massachusetts.

Harvard replied at once to Drew's letter of inquiry. The school would have been glad to admit him—had he only applied earlier. Now, unhappily, it would have to refuse him.

But there was also McGill University in Montreal, the finest of the Canadian schools, on a par with Harvard. McGill also replied without delay, and this time the answer was affirmative. For days Charles Drew walked on air, already imagining himself back in the classroom, beginning the studies that eventually would make him a member of the medical profession.

He came down to earth, though, at the thought of the struggle that lay ahead. To earn a degree at McGill was sufficient challenge in itself. But, leaving home, he would have to earn his living as well. It would be difficult, he concluded, but not impossible. There were summer jobs available, after all, and Drew depended on them for his first two years.

Those were frantically busy years during which Charles Drew applied himself to his studies with a dogged sense of purpose. Even so, he found time to participate in the athletic contests that afforded him so much pleasure. Now, though, he gave up football in favor of high and broad jumping along with hurdling. Once again, he showed that he was championship material by winning the top Canadian trophies.

But 1931, the beginning of Drew's last year at McGill, coincided with the worst period of the worldwide economic depression. There were no jobs available, and it seemed for a while that Charles would not be able to continue school. But at the last minute Drew's university career was saved by a grant from the Rosenwald Foundation. Further help came when he was awarded the Williams Fellowship in Medicine, given to the winner of a competition among the five top students in the class.

The two fellowships enabled Charles Drew to graduate from McGill in 1933 as both a doctor of medicine and a master of surgery. He also had been made a member of Alpha Omega Alpha, the highest honorary society in medicine. With such credentials and his obvious qualifications, he was appointed an intern at Montreal General Hospital and then went on to serve a year as a resident there. In 1935, he was awarded a diploma in surgery from the National Board of Examiners.

Drew was now a doctor and a surgeon. But he had been as fascinated by medical research as he was by the practice of medicine. And his special interest lay in investigations into the problems of blood.

It had first been awakened by one of his professors, a young Englishman named John Beattie. His knowledge was vast, his lectures never confined to the superficial aspects of the matter at hand. When he lectured on bacteriology, he began at the beginning, roamed far and wide, and eventually returned to speak of the most recent winner of the Nobel prize in medicine, Karl Landsteiner, who had been granted the award for his discovery of blood groups. Thirty years had passed before Landsteiner was recognized, Beattie pointed out. But earlier anatomists would have considered themselves incredibly fortunate to be recognized at all. Some, like Servetus, had been burned at the stake. Others, like William Harvey, who discovered the circulation of the blood, had merely been ostracized. All had contributed to the store of human knowledge, but there was still so much more to learn.

Drew had seen, through his experience as a resident, that merely knowing of the existence of different blood groups and the need to match that of the donor to that of the recipient where a transfusion was called for, often meant little in saving lives. Time was also a crucial element. If the proper blood type could be found quickly enough, there was hope for the patient. If not, death was inevitable.

That meant that a supply of blood was essential, too. Precious minutes, even hours, could be wasted in the search for a compatible donor. Surely, he thought, there must be some way to store blood so that it would be available as soon as needed. Perhaps someone had already made experiments in that field, already published the results.

Drew spent hours in the library, searching out all the information he could. He pored over books and scientific papers. He turned the question over and over in his mind.

Drew was preoccupied with the problem of storing blood when he had to face an agonizing personal decision, one common among members of his race who, like him, were fair-skinned, their features more Caucasian than Negroid. His period as a resident was drawing to a close and it was time to

Scientist Versus Society

think of his future. That could be brilliant . . . if he were only white, he was assured by one of his professors, a good friend who spoke to him with the best intentions in the world.

The implication was obvious. Charles Drew could "pass." He could live as a white man, in a white man's world.

But to a person of Drew's integrity, the idea was repugnant. He remembered the example of John Hope, a graduate of Brown University a generation before who had also insisted on guarding his identity as a black. Hope had contributed immeasurably to the progress of his people, first as president of Morehouse College and later as president of Atlanta University. Educators had been needed by Negroes. Doctors, good doctors, were needed, too. No, Charles Drew would not attempt to cross into the white man's world, to live a lie. "That's out of the question," he answered at last. "Absolutely out of the question."

A short time later while Drew was still a resident at Montreal General Hospital, he received an urgent call from Washington, where his father was desperately ill with pneumonia. He left for home immediately, but by the time he arrived the elder Drew was dead.

There was no one now to provide for the family but Charles. Painful though it was, he gave up his residency in Montreal to take a teaching position at Howard University, where he had been rejected as a student only a few years before.

To Drew's dismay, he found that the young men he was to teach were far from dedicated, and his patience wore thin at their indifference. They were slow in learning, not because they lacked the capability but because they lacked the will.

And what motivation did they have, after all? What futures could they look forward to? The hospitals open to them would be wretchedly equipped, often indescribably dirty. Their patients would be the poor, the underprivileged.

True, being a doctor or surgeon, even in a black ghetto, was a certain mark of distinction. To many of Drew's students, it was the sole reason for their choice of career. But they knew, too, that it was a distinction easily achieved.

Because it was inconceivable that black doctors would compete with white ones, the high standards demanded of the latter were often lowered for Negroes. And nowhere was this more true than in the District of Columbia, a federal territory then governed by a congressional committee. Its most prominent members at that time were from the deep American South; some, like Senator Theodore Bilbo of Mississippi, were notorious racists. What they asked of blacks was not excellence in their undertakings but the kind of servility that was a holdover from the days of slavery. Success, in the world they dominated, too often went to the "Uncle Toms." Those who refused to play that role fought a losing battle.

Drew was well aware of the situation, but he vowed that he would never relax his own standards. And so he demanded the best they could give from all whom he taught. To his credit and his satisfaction, he got it. Moreover, he earned their respect. He was turning out doctors who would be a pride to their profession.

In 1938, the Rockefeller Foundation offered a grant that would provide two years' further training at Columbia University in New York to a member of the teaching staff at Howard. The fellowship went to Charles Drew at the end of his third year on the faculty.

He had no hesitation in choosing the field in which he would work. It would be further research on blood, the subject that had fascinated him ever since his days at McGill under Dr. Beattie.

Drew's decision was welcomed by the authorities at

Scientist Versus Society

Columbia, and he was provided with office space as well as a staff of laboratory assistants to carry out the experiments on which he would base a thesis. The successful completion of such a thesis could and did bring him the degree of doctor of medical science and surgery. No other person of his race had ever received it.

Drew's thesis, "Banked Blood," dealt with the problems of storing blood to eliminate the need for direct transfusion from the donor. A method recently developed in Chicago had been a breakthrough, but it had many drawbacks. The system involved refrigerating whole blood, which could be used in place of fresh blood. It was perfectly adequate in most cases. But it also was extremely wasteful since such blood could only be kept for a two-week period. In fact, after one week there was already a noticeable breakdown of the red corpuscles.

Moreover, sodium citrate, the chemical used to prevent clotting, seemed to speed up the destruction of the red blood cells. Blood preserved in such a manner was less satisfactory than fresh blood for the treatment of diseases like anemia.

There were other risks involved, along with the tedious, time-consuming work of typing all blood to be stored. Yet, imperfect as it was, the method filled a desperate need for a ready supply for emergencies. A hospital such as Columbia Presbyterian, where Drew did much of his work, would find a blood bank invaluable. And so he was asked to establish one there on a four-month trial basis.

Charles Drew had been too engaged in his work, too preoccupied with the scientific problems that tantalized him, to devote much time to social pursuits. At the age of thirty-five, his life was practically monastic. Then, on a brief trip to Atlanta with a group of doctors from Freedmen's Hospital, which was associated with the Howard University Medical School, he suddenly fell in love. The young and charming woman, Lenore Robbins, was a teacher at Spellman College for Women in Atlanta. It was a whirlwind courtship—Drew had no time for anything more—and there was no time for a honeymoon when the couple was married in September 1939.

There was added urgency to his work now. The four-month trial period at Columbia Presbyterian had passed, and the emergency blood bank had proved its usefulness. It was not perfect, but important enough so that a permanent place was found for it.

Then war erupted in Europe and it became obvious to Drew that vast quantities of blood would be needed, both for battlefield casualties and for the civilian injuries that would be the inevitable result of air raids. Refrigerated whole blood could never fill the bill. The problem of transportation, to begin with, would be close to insoluble. But that would be almost nothing compared to the problem of matching compatible types.

Without the red corpuscles, though, blood typing wouldn't be needed. Without them—and without the white corpuscles—only the plasma would remain.

Plasma, the fluid part of blood, contained the essential protein, fibrinogen. Lacking that, blood would not coagulate.

But that was the purpose of sodium citrate, which also hastened the destruction of the red corpuscles. Didn't the plasma perform the intrinsic functions he had been searching for? Drew asked himself.

He ticked them off mentally. The loss of protein brought shock in its wake, caused death in the event of serious burns, or death due to loss of blood.

Drew's mind was racing now, posing the critical question. Could plasma replace whole blood?

If it could, the perpetual stumbling block of the disintegration of the red corpuscles would be removed. That in itself would be a tremendous advantage. But there were others—many, many others. Drew now knew where his work lay. Research into plasma. More research. And more.

He at once proposed undertaking the vital task. He would have to add it to his work as director of Columbia Presbyterian's blood bank. And he would need additional funds. But he quickly convinced those who held the pursestrings to grant enough for a six-month project. That six-month period coincided with the last six months of his own fellowship.

While early experiments were promising, promising was not enough. In science, the results had to be conclusive.

But time had run out. Hitler's armies were on the move, and it was evident that France could never stop them.

In April 1940, a meeting was called by the National Blood Transfusion Committee, attended by all experts in the field. France had sent Dr. Alexis Carrel, accompanied by an officer of the French Air Force, to plead for help. The problem was thrashed out in a series of tense meetings that began early in the day and lasted well into the night.

There were objections to the use of plasma; there had been insufficient research on the subject. There was no incontrovertible proof that it would do the job. But there was no alternative, and the Plasma for France Committee was established with Charles Drew as one of the three members. The first shipment was to be sent in six weeks.

It never was.

The French Army had collapsed before the German Juggernaut; the legitimate government had fled. General Philippe Pétain, assuming power, had signed an armistice with the Nazis.

The Plasma for France Committee had no further function and was disbanded. And Charles Drew, his two years of study

Charles Drew

under his Rockefeller grant at an end, returned to Washington, D. C., to his teaching duties and his old post as assistant in surgery at Freedmen's Hospital.

It was now England that bore the brunt of the German attacks.

Wave after wave of German planes swooped across the Channel, raining death and destruction on a defenseless population. The number of victims constantly mounted.

An abundant supply of blood could help save the courageous British people, and so the Blood Transfusion Association was quickly established in New York. It took up the work begun by the Plasma for France Committee.

But the first shipments of plasma turned out to be useless. The blood was contaminated and only sterile blood could be used. Further shipments followed; in some cases, again, the blood was contaminated. It turned out, too, that some hospitals were not filling their promised quotas while others, with more than they could handle, were refusing blood from prospective donors.

Dr. John Beattie, head of the British Transfusion Service and Dr. Drew's old teacher at McGill, soon realized that the trouble stemmed from a lack of organization. He wired New York, urging that a director be appointed to take charge of the whole project. He went so far as to suggest who that should be—his former student, Charles Richard Drew. He knew the field inside and out; it had been his research on which the whole project was based. Besides, he was a skilled administrator. If anyone could get the job done, it was Dr. Charles Richard Drew.

Drew did just that. He set up a central bureau to make sure that facilities existed to accept all donations of blood, ruling out the old hit-and-miss system. He insisted, further, that those who processed the blood should have the highest

Scientist Versus Society

possible credentials. As a final precaution, he arranged that all plasma be checked at Columbia Presbyterian Hospital before it was dispatched to England.

It began to arrive there, sterile, useful—and in the needed amounts. And plasma continued to arrive until 1941, when the British were able to set up their own blood banks.

But Drew's work was far from finished. All signs now pointed to American involvement in the struggle that already engulfed the rest of the world. When it came, another vast store of blood—far greater than any yet envisaged—would be required. The National Research Council, working through the American Red Cross, set up the National Blood Bank Program to deal with the emergency. Thanks to his success with the Blood Transfusion Association, Charles Drew was made its director.

Suddenly, there was an outburst of fury. Blacks were being drafted and were volunteering to fight against the scourge of Nazism, but the blood of their fellow Negroes could not be accepted in that fight. There were, after all, political considerations . . .

Charles Drew knew well what those were. Letters were already pouring in from Southern bigots, objecting to the project in the most violent terms. Death, some insisted, was better than life saved with blood that was other than lily white.

The armed forces themselves were segregated. Blacks did not serve with whites—except, of course, in the traditional role of servant. The situation was explained to Charles Drew by people in authority. He would understand, of course.

But there *was* a solution to the problem, the Red Cross representative insisted. Negro blood could be accepted so long as it was kept separate from that of Caucasians and used only for transfusions to blacks.

Charles Drew

Drew shook his head, concealing his anger only with a mighty effort. Segregation.

"Sit at the back of the bus."

"This drinking fountain for whites only."

"We don't serve you people in this restaurant."

Even when it was a matter of life and death, segregation was the rule of the land—the rule of the country that had included in the preamble to its Constitution the guarantee of "life, liberty, and the pursuit of happiness." But those rights were restricted according to race, now as before, Even when survival was at stake.

Music—the music that his own people had contributed to American culture and that was so widely admired, so widely accepted—was one of Charles Drew's great interests. There had been evenings when he and his friends had gathered around the piano at the home of a colleague to listen for hours to Duke Ellington. At other times, he and his wife had crowded into the small, smoke-filled nightclub where Jelly Roll Morton was playing. Jelly Roll Morton, whose compositions were so original and so important that he had spent days recording them for the archives of the Library of Congress. Drew's thoughts inevitably flashed back to Bessie Smith, the incomparable blues singer.

Bessie, touring the South, had been in a terrible automobile accident. Her body broken and bleeding, she had been rushed to the nearest hospital—and had been turned away! "Whites only!" At a second hospital, the response was the same. Only whites were admitted, but there *was* a hospital for blacks some fifty miles away. The driver turned the car around and raced toward it. When he arrived, Bessie Smith was dead from loss of blood.

Drew's mind snapped back to the present, to the

Scientist Versus Society

patronizing explanations of the Red Cross representative. It was an insult to him and his people. As a scientist, he knew there was no difference between the blood of members of one race and those of another. He could not direct a project that made that distinction.

There was no need even to reflect on the matter. "I'm sorry," Drew said grimly. "I will submit my resignation at once."

Drew called a press conference to explain his actions. He was, he stated, "adamantly, resolutely opposed" to segregating blood. He went on to describe such a policy as "insulting, unscientific, and immoral."

Would the world always be in thrall to "the dark myths and superstitions of the past?" he asked. "Are we the children of bigotry, blinded by yesterday's evil? . . . We share a common blood, but will we ever, ever share a common brotherhood?"

There was a hush in the room as Drew ended, then scattered applause for his integrity and courage. He rose, thanking the reporters, and left the room. In a few hours, he would be on a train for Washington, to again take up his duties at Howard University and Freedmen's Hospital.

He became chief of surgery at the hospital, constantly striving to improve the standards. He devoted himself to teaching, inspiring his students with his own high ideals. And he resigned himself realistically to the prejudice that he must always face, the ever-present humiliations that resulted from discrimination.

Humiliation was frequently the lot of the white people who mixed with blacks in Washington. Public places there were rigorously segregated, at least in one direction. Negroes were never, under any circumstances, allowed in white establishments. Whites, however, were free to go where they wished.

Charles Drew

Black nightclubs were open to them; black schools, too. Even Howard University had a few white students.

Drew, like other doctors at Freedmen's Hospital, could accept white patients as long as they were treated in private offices or at the hospital. On one occasion, a young white woman was referred to Charles Drew by an old friend who happened to be his colleague. She was visiting Washington and had developed a slight abdominal pain, a general queasiness. It hardly seemed worth consulting a doctor; still, it was prudent to mention it to her friend.

A specialist in a completely unrelated field, he hesitated to make a diagnosis himself. But Charley Drew could and was duly called. A quick examination convinced him that the woman had had an attack of appendicitis. He advised an operation, then returned to his other patients.

The woman was prepared to enter Freedmen's, more than willing to have a man of Drew's skill and reputation perform the needed surgery. The second doctor, however, objected. Freedmen's was chronically short of funds for nursing staff, equipment, even for normal maintenance. There were better hospitals in Washington, and they were open to her. "Go to one," he urged. "You're white—you can."

She was appalled at the idea. How could she offend the kind and courteous man who had just left her? "And tell Dr. Drew that a black hospital isn't good enough for me? That a black surgeon isn't good enough?"

"Charley understands."

She shook her head. "No. I can't. I won't."

"I'll tell him, then. Charley understands."

And Charles Drew, devoid of all illusions, did. Fearing certain complications, he even made a small sketch to aid another doctor, in case of emergency.

The drawing was detailed, meticulous. But for all his ex-

Scientist Versus Society

perience of racial prejudice, Dr. Drew had failed to take into account the nature of the complications that were to follow. He had not realized that a white doctor would not accept the patient of one who was black. In the social climate of Washington, D. C., there was only one explanation for the woman's situation. She must have had an illegal operation, an abortion.

It was very late at night before a surgeon was found who was willing to see the woman, considering the circumstances. Even so, it was only under certain specified conditions. She was to be examined in the emergency room of one of the hospitals open to whites, and with three witnesses present.

Drew's diagnosis was, of course, confirmed. A few hours later, a successful operation was performed. Charles Drew was kept informed of the patient's progress, but he never learned of her ordeal. There was no need to inflict still another indignity on top of the many he had already suffered.

The war was over by then. Thousands of lives had been saved by the techniques for storing blood that Drew had developed. More honors had come his way, including the Spingarn Medal, awarded by the National Association for the Advancement of Colored People as the top tribute to one of their race. Well known now, he was offered many chances to enter private practice, where he could considerably increase his income. But he turned them all down. His place was at Howard, he felt, and at Freedmen's, with all its faults. It was only there that he could train other black doctors who would take up his own work after he was gone. He had no idea how soon that would be.

There was to be a conference, one of those held annually, at Tuskegee Institute in Alabama. Drew, who felt he had to go, had planned to make the trip by train. But a group of doctors, also from Freedmen's, were to attend, and it was decided that they would drive there in Drew's car.

Charles Drew

The surgeon was tired as they set off. For a while he drove, then was spelled by another. Late at night, it was Drew's turn again. His head began to nod and he shook it, vainly trying to ward off sleep.

His mind whirled and he gripped the wheel hard, forcing himself to concentrate on the road unwinding ahead of him. But it was useless and his head nodded again as he fell fast asleep.

It was almost dawn then, and the group of doctors had reached the outskirts of Burlington, North Carolina. With Drew asleep at the wheel, the car veered out of control and up an embankment.

The sudden impact startled Drew from his slumber. Aware of the situation now, he fought desperately for control of the machine. As he did, the door beside him opened and he fell part-way out. Summoning all his strength, he struggled to right the car. For a moment it seemed that he would succeed, a moment while it was balanced on two wheels when there seemed to be a chance.

Then the car rolled over on Drew's unprotected body, crushing him beneath a ton or more of steel.

He was rushed to the nearest hospital by his companions, who had escaped almost unscathed. Times had changed since the days of Bessie Smith, and he was admitted at once, despite the color of his skin. But the emergency measures taken to save his life were useless. Charles Drew died on April 1, 1950.

He was forty-five years old.

VII
NICOLAI IVANOVICH VAVILOV

SCIENCE WAS EVERYWHERE on the march in Western and Central Europe throughout the nineteenth century and the first half of the twentieth. In the Eastern part of the Continent, it scarcely lagged behind. In spite of the backward political system of Russia, not far removed from feudalism, the country boasted important universities and distinguished scholars.

The great mass of Russians were illiterate peasants, wresting a difficult living from the land. Only a few in the upper classes were given the privilege of education. But they were well schooled and made major contributions to the natural sciences as well as mathematics.

They were more than willing to accept new ideas, too. When Charles Darwin's theory of evolution shook the world, the initial outcry against it was almost universal. Only in Russia was it fully accepted at once, its ultimate value immediately seen.

Even after the Revolution of 1917, which overturned the tyranny of the czar and replaced his imperial system with the new and never-before-tried ideas of a socialist state, pure science was encouraged. Knowledge was pursued with a real fervor. The new leaders were intellectuals; they understood better than most that progress depended on it.

The upheaval in Russia affected all aspects of life. Education was only one of those. The Communist creed, set forth many years earlier, promised more than the *Liberté, Egalité, Fraternité* of the French Revolution. Instead, it stressed, "To each according to his needs; from each according to his ability."

The Russians, at least at that time, were happy to accept those who had distinguished themselves as scientists before the revolution. One who seemed especially capable, and therefore well worthy of a valued position in the fledgling Socialist Republic, was a noted botanist named Nicolai Ivanovich Vavilov.

He had been born on November 26, 1887, in the capital, Moscow. It was there that he was educated. One subject that fascinated him was the breeding of plants, and he studied the theories of Mendel with great interest and real understanding. His admiration for the Moravian monk was so great that he decided to go to Cambridge University in England to continue his studies. The great attraction of Cambridge was William Bateson, the expert who had written and published a book on Mendel's laws of heredity.

Taking advantage of his stay abroad, Vavilov also worked in London at the John Innes Horticulture Institute. In all, he spent two years studying in Great Britain.

At the outbreak of the First World War, Vavilov returned to Russia, but he went on with his research. Then, as now, the problem of increased production of grain—of wheat, rye, and oats—was pressing. Nowhere was this more so than in the huge, sprawling nation from which he came, where bread was the staple of life. There was little if any industry in the country, and agriculture was the basis of the entire economy. Yet production lagged far behind that of other lands, as the peasants tilled their soil in the primitive style of their forefathers.

There was the climate to contend with, too, the long, bitter winters when the land was frozen. Yet there were means to overcome the problems. For one thing, new and hardier strains might be bred scientifically. To Vavilov, it was evident that such strains might banish the chronic specter of hunger that was so much a part of the life of the Russian peasant.

The more productive and more vigorous species of cereals that Vavilov had in mind were not to be found in his own country. But a number of varieties grew in Persia, not too far away and still accessible. Many of them were superior to the native Russian ones, and so in 1916 Vavilov traveled there, collecting crops for breeding purposes.

When he returned to Russia, his reputation as a man with a sound scientific background, as well as an original thinker, was already established. Consequently, the following year he was made a professor of botany at the University of Saratov, a medium-sized city on the Volga River. He remained there until 1921.

The work he had published both before and after the revolution, as well as his work at the university, put Vavilov in the forefront of Russian scientists. Soon he was appointed to head the All-Union Institute of Plant Breeding. Another and even more significant position followed when Nicolai Vavilov was made director of the Institute of Genetics, a branch of the Russian Academy of Science.

By 1921, Vavilov's work had attracted the attention of Lenin, the first premier of the new socialist state. Setting up the Lenin Academy of Agricultural Scientists, he chose the distinguished botanist to direct it.

Nicolai Ivanovich Vavilov

Vavilov was totally dedicated to his work. Furthermore, he clearly understood the particular problems of agriculture in the Soviet Union. Earlier, he had gone to Persia to seek out special grains. Now, with the backing of the government, he organized a series of expeditions to the places where the species of plants being cultivated for food seemed to have originated. The most competent men he could find were sent to Abyssinia, Afghanistan, South and Central America, and Mexico. In all, there were more than four-hundred such research trips, almost a hundred of them abroad.

The explanation for this worldwide search was found in the so-called principle of diversity, which Vavilov himself expanded and clarified. He had come to the conclusion—one with which most scientists have since agreed—that the plants that are now cultivated had *originated* in the specific regions where the greatest variety of specimens could be found, even though they were often found in other places as well. Cyril Darlington, a noted British botanist, later commented on the Russians's work: "His collections put our ideas of the origins of plants in a new light. They also enabled Soviet breeders to work with the best possible materials in improving the crop plants needed for this new agricultural development of their country."

The expeditions continued until 1933, although a great change had taken place in Russia. In its wake, it brought a new attitude to science.

Nicolai Lenin, the first head of the new Soviet, had died, and a bitter struggle to succeed him had followed. The winner was Josef Stalin, a peasant's son who had risen to the top through his cunning and dishonesty. Within a short time, he was to establish the most brutal of dictatorships, to become as tyrannical as any of the czars.

Stalin understood the practical importance of greater production in agriculture but had no idea of the scientific

Scientist Versus Society

problems involved. Therefore, he pushed for quick solutions. And there were those who promised them.

Among them was Ivan Michurin. To Professor H. J. Muller, a leading American geneticist and a winner of the Nobel Prize in physiology, Michurin was "one of the old practical plant breeders, like Luther Burbank in America." Like the latter, he sometimes succeeded in producing new strains. But, as Muller pointed out, his methods were hardly acceptable to experts, being based "on trial-and-error crossing and rule-of-thumb selection."

Michurin completely ignored the facts about heredity discovered by Mendel and proven true by hundreds who followed him. Moreover, he seemed to have little idea of the real nature of science—the quest for knowledge. In his experiments, he never started out by observation, as Mendel did, drawing his conclusions from what he had observed. In fact, his method was just the opposite; he stated what he thought were certain truths, then searched for evidence to support them.

Muller willingly conceded that some of the varieties Michurin produced, by cross-breeding and plant grafting, were useful. Yet he felt it necessary to add that Michurin's "writings contribute nothing to our understanding of the principles involved."

In Russia, however, Michurin had one powerful and unscrupulous supporter. His name was Trofim Lysenko.

Like Stalin, Lysenko was of peasant birth. And like Stalin, he was a man of overwhelming ambition, coupled with a fanatical streak that pushed him on, destroying anyone in his way, until he attained some of the most important scientific posts in the Soviet Union.

While Muller could grudgingly praise Michurin, he had nothing but contempt for Lysenko. Although he, too, "had achieved some success by the trial-and-error method . . . this gives him no more claim to being a geneticist than does the dubious treatment of a dog for worms." Muller added, "To a scientist, Lysenko's writings along theoretical lines seem the merest drivel. He obviously fails to comprehend either what a controlled experiment is or those known principles of genetics that are taught in any elementary course on the subject."

Lysenko's confusion could be seen in his attitudes toward both Darwin and Mendel. While Mendelian laws were the outcome of Darwin's theories, the Russian pseudo-scientist accepted the idea of natural selection but rejected the means of it. And, with even further confusion, he enthusiastically supported the outmoded ideas of Lamarck. According to Lysenko, it was true that *acquired* characteristics not only *could* be but *would* be inherited. In keeping with this belief, he placed great emphasis on the environment.

Even here, he seemed unable to understand precisely what Lamarck had had in mind. In Lamarck's view, the effect of environment on evolution was limited. Changes would result, he believed, if certain organs were used more frequently. Muscles strengthened by hard work were an example of such use. On the other hand, organs not used would disappear . . . wither away.

One proof of the latter, often cited by Lamarck's followers, was the degeneration or actual blindness of animals living in dark caves. But rigidly controlled experiments with fruit flies, over scores of generations, proved the error of Lamarck's assumptions.

The French scientist's third point was that a specific effort on the part of the plant or animal itself would result in hereditary changes. The giraffe and its long neck, or the webbed feet of ducks, it was assumed, were acquired in this way.

Scientist Versus Society

Today, Lamarck's ideas are regarded as curiosities. No one, however, has discounted certain obvious effects of environment on plants and animals. Studies comparing American children with French children a few years after the Second World War showed that the American youngsters were definitely taller and sturdier, less likely to contract disease, even destined to live longer. The explanation was simple. The American children had been well nourished all their young lives. To them, a morning glass of orange juice was no luxury but a habit. Butter might be in short supply, but there was always plenty of fresh milk to drink. French children in war-torn Europe often had not had enough nutritious food.

The same situation could be applied to agriculture. Plants grown in poor, depleted soil would be shorter and less hardy than those grown in rich loam. The offspring of those plants, grown in the same soil and lacking the same needed nutrients, would again be small. However, once the seeds were sown where there was plenty of sunlight, plenty of water, everything required to nourish them, they would attain their full stature.

It was hardly a question of genetics and had little to do with the remarkable discoveries of Gregor Mendel. The abbot had, after all, used strict controls in his experiments, comparing only plants grown under identical conditions. Lysenko, with no scientific background at all, was muddling the issue rather than clarifying it.

With the same stubborn insistence, and the same lack of understanding, he continued to experiment. It was not long before he announced his discovery of the "vernalization" of plants. This was a means of forcing wheat, normally planted in the winter months, to flower earlier than usual. Lysenko achieved this by soaking the seeds in water, then keeping them at a low temperature for several weeks. He was merely imitating nature, but doing so in his laboratory. There, at least, he was successful. Later, when the same process was applied on a large scale, it failed.

But Lysenko ignored whatever went wrong and made fantastic claims about what he had accomplished. Wheat, so treated, he insisted, would after a few generations flower in just this way, even without forcing. The characteristics acquired over a short period of time would be inherited!

Much work has been done in the field of genetics since Mendel's laws were re-discovered. All of it served to prove Lysenko wrong. Not only that, practical experience reinforced the scientific point of view. Yet most Russian scholars were inclined to be lenient with him.

There was a great need for improvement in agriculture, and Lysenko was working passionately toward that end. While his methods were crude, in the beginning they were useful. Even Nicolai Vavilov supported him at first. Later, when he began to have his doubts, he still gave Lysenko lukewarm acceptance, stating, "He's not doing any harm."

Working at Lysenko's side, and toward the same end, was a philosopher named Prezent, a man thoroughly imbued with Communist party principles. Together, they preached a new and novel line of propaganda. Since, as they insisted, acquired characteristics could be inherited, and since environment was the determining factor for acquiring those characteristics, it was evident that the new Soviet Union would soon raise the miserable, ignorant peasants who made up most of the population to the very highest level. Russians would become supermen.

The two men, posing as respectable scholars, claimed that this was possible. The idea, when reduced to its simplest form, was very close to that of the Communists' political opponents, the Nazis. They, too, believed in a race of supermen, only the

Scientist Versus Society

Nazis insisted that they had always been that chosen group. The Russians, according to Lysenko, could become it.

There were many people, good Communists dedicated to the new system, who thoroughly disagreed with him. Unlike Lysenko, they were all well educated; often they were distinguished scientists, too. They considered the ideas of the peasant farmer to be rubbish.

They began to voice their disapproval, to present opposing views. As they did so, they began to lose their jobs. Soon positions in the universities and the great research centers were closed to Lysenko's opponents.

Deprivation of the right to earn a living was only the first, and the mildest, of the measures taken against these men. It was not long before one after another was arrested by the dreaded secret police. As is typical in totalitarian countries, there would be a knock on the door in the middle of the night. If the door was not opened at once, it would be kicked in. Booted and uniformed agents, well armed, would then ask a few preliminary questions. There would be a search of the apartment, too. After that, Comrade Chetverikov or Comrade Ferry, Comrade Levitsky or Agol would be led off for further questioning.

It would be rare for any to return home. Several great geneticists simply disappeared, their fates unknown to this day. Others were exiled to Siberia, as political opponents of the czars had been.

Conditions there were frightful. The climate was bitter, with temperatures far below freezing for most of the year. Yet those who were detained were denied warm clothing. In addition, they were deprived of the minimum amount of food needed for mere survival. Many died as a result of the terrible hardships they endured in the Arctic wastes. Other geneticists were sent to labor camps, to perform manual labor under conditions scarcely better than slavery. Finally, some were simply

Nicolai Ivanovich Vavilov

liquidated—accused of being enemies of the state, condemned without trial, and shot.

Although this purge of those who supported Mendel's theories began as early as 1932, the world-renowned Medico-Genetical Institute was allowed to function until 1936. It had a staff of many hundreds of superbly gifted and dedicated workers, including biologists, psychologists, and physicians. With adequate financial support from the state, it had been the envy of research centers throughout the world. None had been able to equal its facilities or to attract the uniformly excellent scientists who carried out their experiments there.

The Medico-Genetical Institute was headed by Solomon Levit, who had founded it and built it to its peak of excellence—with government backing, of course. Those who knew the work of the institute knew that the strictest scientific standards were observed. Knowledge could be expanded only by completely objective research; exactly that had been demanded of members of the institute.

In spite of this high ideal, there were a few foreigners who complained that concessions were made to the ideals of communism on some occasions. But this meant nothing when at last the blow fell. Levit became the object of the most insulting, the most violent attacks in the official Soviet newspaper, *Pravda*. Under tremendous pressure from the authorities, he —like Galileo so many centuries before—retracted his earlier words. In a humiliating statement, he admitted his many errors and begged to be forgiven.

Galileo, at least, managed to escape death. Solomon Levit did not. He disappeared soon after his confession; nothing more was ever heard of him.

It was only Vavilov's international reputation that saved him for a few more years. He was, after all, president of the Lenin Academy of Agricultural Sciences, chosen for that post by Lenin himself. He was also to have been president of the

Seventh International Congress of Genetics, the most important meeting in that field in the entire world, scheduled to be held in Moscow in 1937.

But the congress was called off. In its place, a special conference was held in 1936. It was devoted to what amounted to a debate on the merits of the ideas of Mendel and Michurin, as echoed by Lysenko.

It was soon obvious that the entire proceeding had been carefully arranged beforehand. Even the audience, consisting mainly of devoted members of the Communist party who all accepted official doctrine, had been especially selected. As was to be expected in such a case, they applauded almost on signal. And, on signal, they jeered certain speakers, too.

There were many competent, even brilliant scientists on the platform. Although they phrased their remarks with the utmost care, they left no question as to their support of Mendel and Mendelian genetics. But Lysenko also was on hand. He gave the opposite—the Michurin—point of view and was wildly acclaimed.

This was to have been a purely scientific meeting with a discussion of purely scientific matters, yet Lysenko strayed far from them. Instead, he held forth on a number of points of Marxist philosophy. Mendel, he insisted, was an opponent of the Soviets' cherished political doctrine of dialectical materialism. Therefore, Mendel's views were to be dismissed.

The Russian press quickly repeated the accusations against the genuine scientists present. It was then that the attacks on Vavilov began.

The first to be published was written by an agriculturist named Kolj, and it was widely reprinted. Kolj denounced Vavilov for, among other things, having "wasted state money" on "worthless" expeditions to collect plants. He reproved Vavilov for having done little, if anything, to improve crop production. He described him as little more than a charlatan who had deceived the Russian people. Vavilov, who had been called "one of the best scientists Russia ever produced" by the noted British biologist Julian Huxley, was even accused of supporting the racist doctrines of the Nazis, then the archenemies of the Soviets.

It is customary with a scientific conference to publish the proceedings, printing all the papers read. They usually are translated into several languages and distributed to interested and affiliated societies throughout the world.

This was done after the 1936 conference in Moscow. But it was soon evident that many of the speeches defending Mendel and his well-documented work either had been changed or had important passages deleted. A few months later, the Russian government banned even the altered version of the book.

While Vavilov was being attacked by the Soviet press, Lysenko was being extravagantly praised. He became involved in the political life of his country, too, using his growing reputation to further his career. It was through this means that he was able to replace Vavilov as president of the Lenin Academy of Agricultural Sciences in 1938.

A year later, the Seventh International Congress of Genetics, which had been planned for 1936 in Moscow, was at last held in Edinburgh, Scotland. Vavilov, who was to have served as its president, had intended to go. Some forty other Russian scientists had planned to accompany him, and many were to read papers.

At the last minute, however, all were refused permission to leave Russia. At the same time, a letter signed by Vavilov, but obviously written by another person, was received in Edinburgh. With a lack of courtesy that none who knew him could understand, he submitted his resignation as president of the

Scientist Versus Society

congress. It seemed to those who read the note that he had gone out of his way to insult them. Was this from Nicolai Ivanovich Vavilov, gentle and polite, always willing to listen, to discuss at length and with no anger or bitterness any point of view?

Certainly not.

There was a general uneasiness in Edinburgh, an anxiety about the Russian scientist. And yet it seemed that there was nothing to be done. Letters were sent to Vavilov, to be sure. But they were never answered—another strange sign.

In Moscow, however, a conference was being held once again to discuss the respective merits of the two sides of the controversy. But this time not the slightest attempt was made to solve the problems, or even to state them. Those who supported the basic laws of genetics were treated with contempt. The state-controlled newspapers once more began a campaign of hatred and vilification aimed at the genuine scholars of the Soviet Union. Any means was used to disgrace them.

Lysenko's star was rising now. He was soon to be made a vice-president of the Supreme Soviet. Later he would receive the Stalin Prize, one of the highest awards of the U.S.S.R., and he would be awarded it not once but twice. Even greater honors were to come his way, among them the Order of Lenin. And then, most important of all, Lysenko was proclaimed a Hero of the Soviet Union.

But while Lysenko's fortunes were improving, those of Vavilov were growing streadily worse. In 1940, twelve months after the Edinburgh Congress, he was forced from the two distinguished posts he had held for so long, director of the Institute of Plant Industry and director of the Genetics Institute.

Nicolai Ivanovich Vavilov

He had been questioned still another time as to the contributions he could make to Soviet agriculture. More specifically, he had been asked whether or not he could improve plant strains.

Vavilov's answer was, of course, "yes." With his profound understanding of genetics, he didn't hesitate to assure his government that such an undertaking could prove successful. But he had added a word of caution. The development of the new and hardier breeds would take at least five years.

There wasn't time for that, according to the leaders of the U.S.S.R. Others, including Lysenko, were promising immediate results.

Lysenko had only recently presented a new idea and a new technique. Called vegetative hybridization, it consisted of grafting parts of one plant to those of another. Although botanists and biologists have agreed that there is a mutual reaction of the two when this is done, they limit the effect to that. But the new Hero of the Soviet Union proudly proclaimed that the offspring of such a grafted plant would *inherit* the more desirable characteristics of both.

The Second World War was raging over much of Europe by then. Russia had been invaded by the Nazis and was fighting for its very survival. Communications between battle-torn countries were few and far between; little was heard of the controversy over genetics in the Soviet Union, while nothing at all was heard of Vavilov.

But his reputation abroad continually grew. Foreign scientists who had known him mentioned his name and his work in the most complimentary fashion. And nowhere were people more enthusiastic, more impressed by Vavilov than in England. In 1942, his supporters there elected him a foreign member of the Royal Society. There was no higher honor they

Scientist Versus Society

could bestow on him. Only fifty men at any time, from all countries and in every branch of science, can receive such a distinction.

A letter was sent from London to inform Vavilov of the recognition accorded him. But there was no answer.

Further letters were sent, and then the Royal Society began to make inquiries about Nicolai Ivanovich, addressing them to the Soviet Academy of Science.

Again, nothing was heard.

It was not until 1945 that the truth was learned.

In 1940, Nicolai Ivanovich Vavilov, who had done as much if not more for his country than any other man to expand the frontiers of knowledge, had been arrested and accused of being a British spy. That year or the next, he was exiled to Siberia. After a year, he had succumbed to the terrible hardships of life there. By 1942, Nicolai Ivanovich Vavilov was dead.

The wheel had turned full circle.

In the West, governments had at last accepted the challenge of Charles Babbage. They were supporting scientific research, with such success that the new discoveries enabled them to win the war. But in Russia the government had taken complete control of science, subjecting it to political ideas that had no bearing on science.

It had been disastrous to Russian agriculture, which had scarcely advanced beyond its primitive state. The weaknesses of Lysenko's ideas were soon apparent, but it would take years to remedy the damaging effects on all aspects of science.

BIBLIOGRAPHY

Adams, Alexander. *The Eternal Quest.* New York: Putnam, 1969.

Balogh, Penelope. *Freud: A Biographical Introduction.* New York: Charles Scribner's Sons, 1972.

Boas, Franz. *The History of Ideas.* New York: Charles Scribner's Sons, 1969.

Bowden, B. V. *Faster than Thought.* New York: Pitman, 1953.

Buxton, L. H. D. *Charles Babbage and the Difference Engine.* The Newcome Society Transactions, Vol. XIV. London: Courier Press, 1935.

Comrie, L. J. "Babbage's Dream Come True." *Nature*, October 24, 1946, London.

Dampier, W. C. *A History of Science.* New York: Cambridge University Press, 1946.

Darlington, C. D. *The Evolution of Genetic Systems.* New York: Basic Books, 1958.

Darwin, Charles. *Autobiography of Charles Darwin.* New York: Harcourt Brace, 1969.

Bibliography

De Beer, Gavin. *Charles Darwin.* New York: Doubleday & Co., 1964.

Elwin, Malcolm. *Lord Byron's Wife.* London: John Murray, 1962.

Freud, Martin. *Sigmund Freud, Man and Father.* New York: Vanguard Press, 1958.

Freud, Sigmund. *Autobiography.* New York: W. W. Norton & Co., 1935.

———. *An Outline of Psychoanalysis.* New York: W. W. Norton & Co., 1958.

———. *Beyond the Pleasure Principle.* New York: Liveright, 1970.

Halacy, Dan. *Charles Babbage: Father of the Computer.* New York: Crowell-Collier Press, 1970.

Hardwick, Richard. *Charles Richard Drew: Pioneer in Blood Research.* New York: Charles Scribner's Sons, 1967.

Huxley, Julian. *Heredity East and West.* New York: Henry Schuman, 1949.

Iltis, Hugo. *Life of Mendel.* London: George Allen & Unwin, Ltd., 1932.

Irvine, William. *Apes, Angels and Victorians,* New York: McGraw-Hill, 1955.

Jones, Ernest. *Sigmund Freud, Life and Works.* London: Hogarth Press, 1958.

Karp, Walter and Burrow, J. W. *Charles Darwin and The Origin of Species.* New York: American Heritage, 1968.

Langley-Moore, Doris. *The Late Lord Byron.* London: John Murray, 1941.

Levine, Lawrence W. *Defender of the Faith.* Fairlawn, N.J.: Oxford University Press, 1965.

Lichello, Robert. *Pioneer in Blood Plasma.* New York: Julian Messner, 1968.

Logan, Rayford W. *The Betrayal of the Negro.* New York: Collier Books, 1965.

Bibliography

Mason, S. F. *Main Currents of Scientific Thought.* New York: Abelard-Schuman, 1956.

Maynes, Ethel Colburn. *The Life and Letters of Anne Isabella, Lady Byron.* London: Constable, 1939.

Medvedev, Zhores. *The Rise and Fall of T. D. Lysenko.* New York: Columbia University Press, 1969.

Moseley, Maboth. *Irascible Genius: The Life of Charles Babbage.* Clifton, N.J.: Kelley, 1964.

Morrison, Philip and Emily. *Charles Babbage and His Calculating Engines.* New York: Dover Publications, Inc., 1961.

———. "Charles Babbage." *Scientific American,* April, 1952.

Muller, H. J. "The Destruction of Science in the USSR." *Saturday Review of Literature,* December 4 and December 11, 1948.

Mullett, Charles F. "Charles Babbage, A Scientific Gadfly." *Scientific Monthly,* London, November, 1948.

Purver, Margery. *The Royal Society: Concept and Creation.* Boston: Routledge & Kegan Paul, 1967.

Richardson, Ben. *Great American Negroes.* New York: Thomas Crowell, 1956.

Sears, Paul B. *Charles Darwin.* New York: Charles Scribner's Sons, 1950.

Stern, Curt and Sherwood, Eva R., eds. *The Origin of Genetics.* San Francisco: W. H. Freeman.

Sterne, Emma Gelders. *Blood Brothers: Four Men of Science.* New York: Alfred Knopf, 1959.

Strickland, Margot, *The Byron Women.* London: Peter Owen, 1974.

Tenbroek, Jacobus. *Equal Under Law.* New York: Collier Books, 1965.

Turney, Catherine. *Byron's Daughter.* London: Peter Davies, 1974.

Webster, Gary. *The Man Who Found Out Why.* London: Kingswood & Co., 1968.

INDEX

Académie des Sciences (France), 19-20
Academy of Science (U.S.S.R.), 138, 150
Agassiz, Louis, 16
Agricultural Society of Moravia and Silesia, 74
Airy, George, 29, 30, 33
Albert, Prince, 36
Algebraic system, for computers, 34
All-Union Institute of Plant Breeding (U.S.S.R.), 138
Alpha Omega Alpha (medical society), 122
American Red Cross, 130
Amherst College, 118-120, 121
Analytical Engine (computer), 32-37, 40, 41, 54, 55-56
   design for, 32-33
   development of, 34-35
Analytical Society, 21, 24
Anti-Semitism, 89
Archimedes, 37
Atlanta University, 124

Babbage, Benjamin, 24, 26, 29
Babbage, Charles, 18, 19-41, 42, 48, 51, 53, 54, 56, 150
   and Analytical Engine concept, 32-37, 40, 41, 54, 55-56
   birth of, 21
   death of, 40
   and Difference Engine concept, 26-32, 33, 40-41, 51, 55
   education of, 21-24
   friendship with Countess Lovelace, 54, 55, 58-59
   inventions and discoveries of, 34-35
Babbage, Georgiane, 24, 26, 29
Bateson, William, 86, 137
Beattie, John, 122, 123, 125, 129
Bernays, Martha, 91, 94, 95-96
Bernouilli, Jacques, 55
Beyond the Pleasure Principle (Freud), 105, 106
Bible, 16
Bibliothèque Universelle de Genève, 37

Index

Bilbo, Theodore, 125
Binary system of mathematical notation, 32
Blood Transfusion Association, 129, 130
Bonaparte, Princess Marie, 105, 112
Breuer, Josef, 92-93, 97, 98, 100
British Post Office, 34-35
British Transfusion Service, 129
Brown v. *Board of Education of Topeka*, 114
Browning, Robert, 101
Brücke, Ernst, 91
Brünn Modern School, 72
Brünn Society for Natural History, 84
Brünn Society for the Study of Natural Science, 74, 79-80
Brünn Theological College, 69
Buddhists, 14
Bullitt, William, 111
Burbank, Luther, 140
Byron, Ada Augusta, *see* Lovelace, Ada Augusta, Countess of
Byron, Fifth Lord, 42
Byron, George Gordon, Lord, 37, 42-47, 48, 50, 57-58
Byron, John "Mad Jack," 43
Byron, Lady (Anne Isabella Milbanke), 43-47, 48, 50-51, 54, 56, 59, 60

Calculators. *See* Difference Engine; Analytical Engine
Cambridge University, 20, 21-24, 30, 53, 56, 137
Camerarius, Dr. Joachim, 77
Carrel, Dr. Alexis, 128
Charcot, Jean Martin, 93, 94-95
Charles Albert, King, 36, 37
Charles II, King, 19
Childe Harold (Byron), 43
Church of England, 15
Civilization and Its Discontents (Freud), 109

Clark University, 105
Cobb Pentathlon Award, 118, 119
Cocaine, 94
Colbert, Jean Baptiste, 19
Columbia Presbyterian Hospital, 126, 128, 130
Columbia University, 125-127
Comrie, L. J., 40
Constant difference, 27
Copernicus, Nicolaus, 14-15
Correns, Karl Erich, 84

Darlington, Cyril, 139
Darwin, Charles, 11-12, 16-18, 74, 75, 81, 83, 101, 136, 141
Darwin, Erasmus, 12, 74
Davy, Humphrey, 38
Death instinct, theory of, 105-106
Deism, 25
Depression of 1930's, 122
Descartes, René, 19
Difference Engine (computer), 26-32, 33, 40-41, 51, 55
  construction of, 30-32
  cost of, 30
  working model of, 26-28
Differences, principle of, 26-28
Differential and Integral Calculus (Lacroix), 25
Disraeli, Benjamin, 35-36
Diversity, principle of, 139
Drew, Charles, 18, 114-135
  and "Banked Blood" thesis, 126
  birth of, 114
  college athletic career of, 117-120
  death of, 135
  education of, 115-116, 121-122
  and experience of racial prejudice, 130-134
  internship of, 122-124
  music interests of, 131
  sports awards of, 118, 119, 120
  and techniques for storing blood, 126, 127-128

156

Index

Drew, Lenore, 127
Dunbar, Paul Laurence, 115

East India College, 25
Economy of Manufacturers and Machinery, The (Babbage), 34
Ego and the Id, The (Freud), 107
Ellington, Duke, 131
Evolution, theory of, 11–12, 17–18, 80–81, 83, 101, 136

Ferenczi, Sandor, 105, 109
Fibrinogen, 127
Focke, W. O., 83–84
Fraser, Donald, 103
Freedmen's Hospital, 126, 129, 132, 133, 134
French air force, 128
French Revolution, 20, 137
Freud, Amalie, 88–89
Freud, Anna, 109, 110, 111, 112
Freud, Emmanuel, 87
Freud, Jakob, 87, 88, 89
Freud, John, 87
Freud, Martha, 107
Freud, Martin, 111
Freud, Sigmund, 18, 87–113
  awarded Goethe prize (1930), 110
  birth of, 87–88
  death of, 112–113
  education of, 89–90
  medical operations of, 108
  Nazi terror and, 110–112
  in Paris, 93–95
  psychoanalytic movement and, 98–106
  and study of nervous diseases, 92–93
  and technique of free association, 97–98
  use of cocaine by, 94
  use of hypnosis by, 96–97
Freud, Sophie, 107
Freund, Anton von, 107

Galileo, 15, 37, 145
Genetics Institute (U.S.S.R.), 138, 148
Ghost Club, 24
Goethe, Johann Wolfgang von, 90, 110
Gravity, principle of, 25
Group Psychology (Freud), 107

Harvard University, 121
Harvey, William, 123
Henry VIII, King, 42
Herschel, John, 23, 25
Hitler, Adolf, 108, 110–112, 128
Hope, John, 124
Hôpital Salpêtrière, 93
Howard University, 115–116, 120, 121, 124–125, 132, 134
Howard University Medical School, 126
Huxley, Julian, 147

Iltis, Hugo, 85
Index of Roman Catholic Church, 74
Institute of Plant Industry (U.S.S.R.), 148
Interpretation of Dreams, The (Freud), 99

Jacquard, Joseph, 32
Jacquard-loom, 55
John Innes Horticulture Institute, 137
Jones, Ernest, 103, 107
Judaism, 14
Jung, C. G., 104–105

Kepler, Johannes, 15
King, William, 49
Koniginkloster (Augustinian monastery), 68–69, 73, 82

Lacroix, Silvestre François, 25
Lamarck, Jean Baptiste, Chevalier de, 12, 141, 142

157

Index

Lamarckian theory, 18
Landsteiner, Karl, 123
Leibnitz, Gottfried Wilhelm von, 23, 25, 26
Leigh, Augusta, 47, 58
Lenin, Nikolai, 138, 139, 145
Lenin Academy of Agricultural Sciences, 138, 145, 147
Levit, Solomon, 145
"Life tables," insurance policy, 35
London *Times*, 39, 40
Lovelace, Ada Augusta, Countess of, 18, 37, 42-60
  birth of, 46
  death of, 59-60
  education of, 49-50, 52
  and friendship with Babbage, 54, 55, 58-59
  horseracing interests of, 51-52, 58-59
  and translation of Menebrea's book, 54-56
Lovelace, Byron, 53, 58
Lovelace, First Earl of (William Lord King), 53
Lovelace, Ralph, 58
Luther, Martin, 15
Lysenko, Trofim, 140-149, 150
  awarded Stalin Prize, 148
  opposition to theories of, 144-145
  vegetative hybridization technique of, 149

McGill University, 121-122, 129
Mary Stuart, Queen, 42
Mechanics Institute (London), 51, 53
Medico-Genetical Institute (U.S.S.R.), 145
Medora, Elizabeth, 47, 58
Mein Kampf (Hitler), 108
Mendel, Anton, 62, 63-64, 65, 66, 67
Mendel, Gregor, 18, 61-68, 137, 141, 142, 146
  birth of, 62-63
  botany experiments of, 75-81
  death of, 83
  and decision for the priesthood, 68-69
  education of, 64-67, 71
  religious training of, 69-70
  and tax dispute with the government, 82-83
  teaching career of, 70, 72, 73
Mendel, Rosine, 62
Mendel, Theresia, 67, 82
Mendel, Veronika, 66
Mendel's law of heredity, 79, 84, 85, 137, 140, 141, 143, 145, 146
Mendel's Principles of Heredity (Bateson), 86
Menebrea, L. F., 37, 54
Meynart, Theodor, 92
Michurin, Ivan, 140, 146
Milbanke, Lady, 46, 49, 54
Monotheism, 14
Montreal General Hospital, 122, 124
Moravian Institute for Deaf Mutes, 83
Moravian Mortgage Bank, 83
Morehouse College, 124
Morgan, Augustus de, 54, 56-57
Morgan College, 121
Morton, Jelly Roll, 131
Moses and Monotheism (Freud), 112
Moslems, 14
Mozart, Wolfgang Amadeus, 50
Muller, H. J., 140

Nägeli, Karl Wilhelm von, 81
Napoleon I, Emperor, 62
National Blood Bank Program, 130
National Blood Transfusion Committee, 128
National Board of Examiners (Canada), 122
National Research Council, 130

Index

Nazi Party, 105, 108, 110–112, 128, 130, 143, 147, 149
Newton, Isaac, 15, 21, 25, 30
Nobel, Alfred, 101

Oedipus complex, 99
Operational research methods, 34
Origin of Species by Means of Natural Selection, The (Darwin), 11–12, 16–17
Outline of Psychoanalysis, An (Freud), 112
Oxford University, 20, 53

Pascal, Blaise, 19, 23, 26
Peacock, George, 23, 25
Peel, Robert, 33, 34
"Penny post" mail system, 34–35
Pétain, Philippe, 128
Plana, Baron, 36
Plasma for France Committee, 128–129
Pravda, 145

Rank, Otto, 109
Reflections on the Decline of Science in England and Some of Its Causes (Babbage), 31
Religion, 12–16
  early forms of, 12–13
Rh blood factor, 85
Robeson, Paul, 116
Rockefeller Foundation, 125
Roman Catholic Church, 14, 16, 63, 74
Roosevelt, Franklin D., 111
Rosenwald Foundation, 122
Royal Astronomical Society, 28, 33
Royal Society (Great Britain), 19, 24, 31, 38, 53, 112, 149–150
Russian revolution of 1917, 136–137
Rutgers University, 116

Schur, Max, 112
Scopes, John Thomas, 17
Seventh International Congress of Genetics, 146, 147–148
Servetus, Michael, 15, 123
Sheepshank, Richard, 33, 34
Sketch of the Analytical Engine Invented by Charles Babbage, A (Menebrea), 37
Smith, Bessie, 131, 135
Sodium citrate, 126
Somerville, Mrs. William, 53
Spellman College for Women, 127
Stalin, Josef, 139–140
Studies on Hysteria (Freud and Breuer), 97, 98

Tennyson, Alfred Lord, 39
Thaler, Aurelius, 69
Three Essays on the Theory of Sexuality (Freud), 100
Totem and Taboo (Freud), 105
Tschermak, Erich von, 84
Tuskegee Institute, 134

University of Edinburgh, 26
University of London, 110
University of Saratov, 138
University of Vienna, 71, 102, 107

Vavilov, Nicolai Ivanovich, 136–150
  birth of, 137
  death of, 150
  elected to Royal Society, 149–150
  plant research of, 137–138, 139, 140
  Siberian exile of, 150
Verne, Jules, 35
Victor Emmanuel II, King, 37
Victoria, Queen, 36
"Vision of Sin, The" (Tennyson), 39
Voltaire, 101
Vries, Hugo de, 83–84

Index

Waldberg, Count, 62
Waldberg, Countess, 63
Werner Society, 74
Whist Club, 24
Whitmore family, 24
Williams Fellowship in Medicine, 122

World War I, 106, 137
World War II, 101, 127, 128-129, 134, 149

Zoological and Botanical Society of Vienna, 71